Praise for *Hope Beneath Our Feet*

"Uplifting and diverse perspectives on aligning ourselves with the fullness of our own possibilities as individuals and as a species. The bottom line: flavor and savor every moment, especially the toughest ones, with not-knowing, with good will, imagination, kindness, wisdom, humor, community, and action. My father-in-law, Howard Zinn, said it so well: 'Small acts, when multiplied by millions of people, can transform the world.'"

– Jon Kabat-Zinn, author of *Full Catastrophe Living* and *Coming to Our Senses*

"*Hope Beneath Our Feet* is just that. With penetrating clarity it takes us into the depths of our essential nature: our courage, brilliance, and resilience. We are called to take advantage of the greatest opportunity for conscious self-evolution the human species has ever been given. A real masterwork."

–Barbara Marx Hubbard, author of *Conscious Evolution: Awakening the Power of Our Social Potential* and founder of the Foundation for Conscious Evolution

"There is no more meaningful action we can take than to struggle on behalf of life on this planet. The profoundly hopeful and inspiring essays in this anthology help us remember our place in the web of life and recover a deep awareness of our own ecological identity, motivating us to continue our efforts."

–John Seed, founder of the Rainforest Information Centre in New South Wales, Australia, and co-author of *Thinking Like a Mountain: Towards a Council of All Beings*

"This fierce collection is a pragmatic and poetic call to action for the environmental predicament that we've authored. Its message is compelling and, if acted upon, contains powerful medicine to heal ourselves and the planet."

–Richard Strozzi-Heckler, author of *The Leadership Dojo*

"A major antidote for despair is engagement and participation: that is the key message of this collection of leading thinkers and doers hard at work on the dire problem of worldwide ecological decline. What creative responses to the crisis of our time await beyond weary save-the-world technologism, activism, and heroism? What can *you* do, today, to help? Find out here."

–Craig Chalquist, co-editor of *Ecotherapy: Healing with Nature in Mind* and core faculty member at John F. Kennedy University

Hope Beneath Our Feet

Restoring Our Place in the Natural World

AN ANTHOLOGY

Edited by
Martin Keogh

North Atlantic Books
Berkeley, California

Published by
North Atlantic Books
P.O. Box 12327
Berkeley, California 94712

Cover and book design © Ayelet Maida, A/M Studios
Cover images © Eric Isselée (fish), © Pavel Lebedinsky (bee) / iStockphoto.com
Printed in the United States of America

This is issue number 67 in the Io series.

Hope Beneath Our Feet: Restoring Our Place in the Natural World is sponsored by the Society for the Study of Native Arts and Sciences, a nonprofit educational corporation whose goals are to develop an educational and cross-cultural perspective linking various scientific, social, and artistic fields; to nurture a holistic view of arts, sciences, humanities, and healing; and to publish and distribute literature on the relationship of mind, body, and nature.

North Atlantic Books' publications are available through most bookstores. For further information, visit our Web site at www.northatlanticbooks.com or call 800-733-3000.

Copyright acknowledgments are listed on page 304. Every effort was made to contact the holders of copyright to material reprinted in this book.

Library of Congress Cataloging-in-Publication Data

Hope beneath our feet : restoring our place in the natural world : an anthology / edited by Martin Keogh.
     p. cm.
   Summary: "Hope Beneath Our Feet is an anthology of essays that answer the question, 'If we are facing imminent environmental catastrophe, how do I live my life right now?'"—Provided by publisher.
   ISBN 978-1-55643-919-3
1. Sustainable living. 2. Environmentalism. I. Keogh, Martin J., 1958–
   GE196.H67 2010
   304.2—dc22
                                                  2010014998

2 3 4 5 6 7 8 9 SHERIDAN 15 14 13 12 11 10

Contents

Introduction by Martin Keogh – ix

CHAPTER ONE: *What's at Stake*

Commencement Address – Paul Hawken – 3
Awakening to Our Evolutionary Responsibility
    – Andrew Cohen – 8
The Original Human Vocation – Barbara Kingsolver – 11
Living with Losing – Ben Gadd – 15
Humanity's Rite of Passage – Anodea Judith, PhD – 24
Why Bother? – Michael Pollan – 29

CHAPTER TWO: *A Way Forward*

Letter from the Future – Vicki Robin – 41
A Way Forward in an Uncertain Future
    – Susan Feathers – 49
Living with Purpose in the End Times – Jamie McHugh – 53
Love the Things We Love – Larry Santoyo – 58
Gandhi Then and Now – Michael N. Nagler – 62
Inspiring and Sustaining Action Over Time
    – Ruskin K. Hartley – 68

CHAPTER THREE: *Taking Single Steps*

Dusting Off the Energy Solution in the Basement
    – Dana Goldman – 73
Every Day We Choose – Frances Moore Lappé – 76
With the Turn of a Key, I Can Make a Difference
    – Neige Christenson – 82
Shut Up and Vote – Eric Rubury – 86
One Piece of Paper – Kristine Alach – 90
A Five-Hundred-Year Plan – Jane Hayes – 93

CHAPTER FOUR: *Little Steps to Big Leaps*

Fight It Head On – Bill McKibben – 99

Become an Urban Homesteader – Kelly Coyne
 and Erik Knutzen – 102

To Build a Better Future, Start with a Better Question
 – Jeffrey Hollender – 106

Getting Ready for Change – Bill Mollison – 111

Thinking Like an Island – Michael Ableman – 115

The World Is Falling Apart! What Should I Do?
 – Susan Bartlett – 122

Challenging a Corporation to Clean Up Its Act
 – Thaïs Mazur – 125

Nothing Else Matters – Derrick Jensen – 132

CHAPTER FIVE: *The Body of the World*

Eyes Wide Open – Chameli Gad Ardagh – 137

The Healing Power of Nature – Diane Ackerman – 140

A Sense of Place—A Sense of Self – Ian McCallum – 145

Body as Place – Nala Walla – 150

Morality Is a Somatic Experience – Tom Myers – 158

Earth Rights – Dr. Vandana Shiva – 166

Indigenous Mind – Kaylynn Sullivan TwoTrees – 170

CHAPTER SIX: *Balanced Engagement*

Wonder: A Practice for Everyday Life
 – Munju Ravindra – 177

Embodying Change – Cheryl Pallant – 183

The World Doesn't Need to Be Saved – Byron Katie – 188

What Keeps Me Alive: Making It Real – Chaia Heller – 193

In the Climate Era the Personal Is Political
 – Tzeporah Berman – 197

Coping with New Realities – Linda Buzzell – 201

CHAPTER SEVEN: *Meditations on Living in These Times*

    Eden Is a Conversation – Barry Lopez – 209

    Fostering Light in Dark Times – Vivienne Simon – 213

    From Mourning into Daybreak – Nina Simons – 217

    Waking Up from Despair – Opeyemi Parham – 225

    River Gods – Ken Lamberton – 232

    Questions for a Sacred Life – Bodhi Be – 238

    To Do the Will of God, Come What May
       – Alice Walker – 243

CHAPTER EIGHT: *Hope in Challenging Times*

    To Endure Climate Chaos, Live Dangerously
       and Cultivate Hope – Brian Tokar – 247

    Fighting Fatalism about War – John Horgan – 253

    Little by Little – Margaret Trost – 256

    The Grandmothers Speak – Jeneane Prevatt
       and Ann Rosencranz – 260

    The Ultimate Miracle Worker – Jalaja Bonheim – 265

    The Challenge of Building Sustainably – Scott Rodwin – 269

    The Optimism of Uncertainty – Howard Zinn – 277

AFTERWORD

    Sabbaths: VI – Wendell Berry – 283

    Gratitude – 287

    Index – 291

    About the Editor – 303

    Permissions and Copyrights – 304

Introduction

MARTIN KEOGH

Sometimes bumping into a single piece of information can wake a person up to the plight of our world. This awareness came to me a few months after my son Dylan was born. In the warm comfort of our living room on a New England winter evening, I sat reading statistics on the decline of the world's coral reefs. Glancing over at the face of my infant son as he slept in his mother's arms, I imagined the world that he is to inherit. Those dying reefs suddenly did not feel far away—or so far in the future.

I was stunned to learn that, while estimates differ, in a few decades—or maybe even less time—the coral reefs could be virtually gone. Coral reefs are a life-support system not only for themselves and countless fish species, but also for the three hundred million people whose sustenance depends on the seafood harvested in these waters. We will not only have to cope with the loss of an entire habitat teeming with life—we will be staring right in the face of a global food-source collapse.

Events that many of us imagined would not threaten children until future generations are occurring even as we sit down to our dinner.

This recognition ushered in a series of sleepless nights. I lay awake as images crowded my mind: the seas filled with more specks of plastic than krill; axes and torches leaving stumps as they progressed through the Amazon rainforest, our "lungs of the earth"; and much closer to home, fewer and fewer songbirds on the branches of our own neighborhood trees. Grieving over all this loss, I wondered, *If what is happening is so utterly different than anything we've experienced in our lifetimes, how do I live in the face of such loss?*

My wakeful nights did not serve my family, or the world. My heart would pound so hard in my chest that my wife Liza would feel it through the mattress and then she too would lie awake.

In the midst of this despair, a friend emailed me "The Peace of Wild Things" by Wendell Berry:

> When despair for the world grows in me
> and I wake in the night at the least sound
> in fear of what my life and my children's lives may be,
> I go and lie down where the wood drake
> rests in his beauty on the water, and the heron feeds.
> I come into the peace of wild things
> who do not tax their lives with forethought
> of grief. I come into the presence of still water,
> and I feel above me the day-blind stars
> waiting for their light. For a time
> I rest in the grace of the world, and am free.

I took the poem's advice to heart. When I found myself awake, I would walk to the end of our street and continue into the woods. Sometimes the trail would be lit only by starlight. Feeling my body surrounded by so much life calmed my galloping heart.

One morning, as dawn arrived, I strolled into a nearby meadow. With each step rose a blur of hundreds of Ruby Meadowhawk dragonflies, which settled back down only to rise again as I took another step. As I watched the alarmed flapping of one of these graceful creatures escaping my shoe, I realized that I was not big enough to hold these questions alone, that I needed the help of others.

So I asked people how they were coping with the ongoing flood of news from the receding edges of nature. The more people I asked, the more I realized that I was not alone in asking. During my travels to teach in different parts of the world, I increasingly heard similar questions. A woman in Helsinki phrased it this way:

If our world is really looking down the barrel of an envi-
ronmental catastrophe, how do I live my life right now?

One woman responded to this question by fluttering her hands (a flutter that I now recognize goes with issues too big to imagine): "With so much stress in our everyday lives, how can we think about *that*?"

Some people were adept at changing the subject. Often they brought up another problem: "My husband never gets home when he says he will" … "The clutch on my Chevy keeps giving out" … "The IRS is auditing my taxes again." They shifted the problem to a scale that was imaginable, one that they could wrap their heads around.

When people respond to these questions, I sometimes hear despair. More than once, I was told, "We are just rearranging the deck chairs on the Titanic." The desire for practical suggestions is strong for many people; they want to know what they can *do*. Others expressed a longing for a spiritual perspective.

My inquiries made me realize the urgent need for thoughtful people to provide reflection, inspiration, and direction. So I sent the question of the Finnish woman to environmentalists, artists, CEOs, grassroots organizers, religious and indigenous leaders, scientists, and folks who simply are concerned, to hear how they would respond.

The first wave of replies came from people who feel resigned to imminent disaster. One wrote of friends who are stocking caves in the Sierra; another person grows potatoes on his rooftop as emergency food. One person likened the inflation of our planet's population to a stock market bubble, due for a big "correction"—and soon. Several of these responses came with the addendum, "Hug your loved ones while you still have the chance."

And then responses trickled in from people taking time in their lives to seek remedies. Some were gathering with their neighbors to build sustainability groups; others had started grass roots organizations; still others were negotiating a closer marriage between science and public policy. Friends told friends about my query, and soon a flood of people who had meditated on these questions were offering to contribute to this anthology.

I began to sleep better, knowing that so many people care and are taking a stand in their communities for what they feel and know in their hearts. These individuals helped me discover that a major antidote for despair is engagement and participation. Their responses to the question "How do I live my life right now?" have been compiled into the anthology you now hold in your hands: *Hope Beneath Our Feet*.

We Have a Choice

Al Gore's movie, *An Inconvenient Truth,* demonstrated how the misconceptions and misdeeds of our civilization have put us in a dire predicament. This film clearly shows that we each have a choice: either we find effective ways to contribute to making changes now, or we will have to sacrifice—with unbearable losses—later.

My family watched the film together. A few days later, Liza and I asked our two teenagers for their reaction. There was a long silence, and then William summed up the movie: "So basically, we're fucked, right?"

After seeing the film, many people commented that they now needed tools, both practical and spiritual, to handle this new awareness. They wanted more. I could tell that William and his brother Wyatt had been jarred, but they never brought the movie up again. Even so, things started to change in our home. The boys remembered to turn off the lights in rooms that weren't occupied. They griped less about taking out the compost and separating the recycling. They surprised us by choosing to eat less red meat. And they became interested in the efforts that Liza and I had been making for three years to lower our carbon footprint.

Walking into the kitchen, I would find them with their friends, assembled around the refrigerator, munching on an after-school snack and peering at the sheet we'd put up to show how actions such as installing energy-efficient appliances, taking shorter showers, and buying a hybrid car had considerably reduced our resource consumption. Water and heating oil: down by 35 percent. Gasoline: cut by 80 percent. My airplane business travel: down 50 percent. What surprised our boys was how little noticeable sacrifice had to be made to produce these savings (though they did miss our old minivan).

However, when we suggested that they join a sustainability group at school, Wyatt dismissed the idea: "Nope, only the hippies do that." But when our family bought a share in a local CSA (Community-Supported Agriculture) farm and began to receive a box each week that spilled over with organic produce, the boys held their noses and tried some vegetables they had never set eyes on before. Some lived up to their expectations and others they found surprisingly tasty, especially when cooked in the solar oven.

No one really knows whether this one teenager's comment after watching *An Inconvenient Truth* is accurate. The truth is, we might have arrived at the tipping point. What the contributors in this book reveal, however, is that if enough of us lean together in the right direction, our trajectory can change; we *do* have the ability to alter the course of events. We *have* to make this effort—because the alternative is unthinkable.

The only way we are going to make it is if everyone contributes to the many remedies needed. And the good news is that the world has never given us such a vital opportunity both to find our contribution and to offer it.

Restoring Our Place in the Natural World

We need only turn on the evening news to hear the litany of what is wrong around us. In these essays and meditations, you will not find a catalog of despair. This book is an invitation to move beyond merely coping into actively engaging.

When I sent out the initial requests for writings, I did not know what form the book would take. I didn't know that it would become an invitation, a challenge, a spur, for each reader—for you—to find his or her own particular ways to contribute. The authors describe myriad approaches to finding the drive and passion and will to stand up for our world. Through their eyes, we discover that our solutions are as multifaceted as our problems, making room for each of us to weigh in with our own style. Your approach may be through science, advocating for legislation, chaining yourself to a tree, or simply starting conversations. You might have a skill that can support the good work of others. Most likely, your part will include simply lowering your own consumption of our earth's resources.

Ideally you will hear more than a few voices within these pages that speak directly to you. I invite you to seek them out.

Kelly Coyne and Erik Knutzen show simple ways to make a difference, literally in your own backyard. Jeneane Prevatt and Ann Rosencranz connect us to an indigenous wisdom with our feet in the natural world. Several authors, including Diane Ackerman and Alice Walker, demonstrate that gratitude can lead to forms of activism. Frances Moore Lappé, John Horgan, and Margaret Trost reveal how your actions indeed send out ripples that have influence. And Derrick Jensen lets us know, in no uncertain terms, why there is no time to wait.

Not everybody here agrees with each other, nor should they. But their generosity is born of a passion to see all of us meet the challenge—together. Some have been laboring for decades, trying to wake people up to the reality of what we are doing to our world and ourselves. They grasp the shocking details of our situation—yet for the most part these people are filled with joy, if not hope, for the ingenuity and initiative that people are capable of. These individuals are not living in isolation; they've learned to balance struggle with celebration.

It's worth saying again: they have found the antidote for despair in participation.

The authors' work "on the ground" gives testimony that responsible engagement reconnects us to the world of which we are a part. In this book you will also encounter a Malaysian hairy rhino, many birds, including red tail hawks, trogons, and rose-throated becards. You will find green rat snakes, gila monsters, javelinas, and spider monkeys. We can thrive on the body of this earth only when we stop seeing the earth and its inhabitants as separate from ourselves and our survival. Every living being is part of the remedy.

None of us knows for sure which side of the tipping point we are on. But I imagine that you share with me the desire to look back at the end of our lives and feel that we have lived each moment fully engaged, knowing that we've each contributed our small share. To do this, we need the humility to recognize that we are not going to figure this out alone. So much of what we face is unfathomable. We need to develop the capacity to reach out to one another, and to call on something intangible beyond ourselves.

I invite you to read on—and to create your own *Hope Beneath Our Feet* project. In ways you cannot even imagine, what you do matters and makes a difference for us all.

CHAPTER ONE

What's at Stake

Commencement Address to the Class of 2009, University of Portland

PAUL HAWKEN

When I was invited to give this speech, I was asked if I could give a simple, short talk that was "direct, naked, taut, honest, passionate, lean, shivering, startling, and graceful." No pressure there.

Let's begin with the startling part. Class of 2009: you are going to have to figure out what it means to be a human being on earth at a time when every living system is declining, and the rate of decline is accelerating. Kind of a mind-boggling situation ... but not one peer-reviewed paper published in the last thirty years can refute that statement. Basically, civilization needs a new operating system, you are the programmers, and we need it within a few decades.

This planet came with a set of instructions, but we seem to have misplaced them. Important rules—like don't poison the water, soil, or air; don't let the earth get overcrowded; and don't touch the thermostat—have been broken. Buckminster Fuller said that "spaceship earth" was so ingeniously designed that no one has a clue that we are on one, flying through the universe at a million miles per hour, with no need for seatbelts, lots of room in coach, and really good food—but all that is changing.

There is invisible writing on the back of the diploma you will receive, and in case you didn't bring lemon juice to decode it, I can tell you what it says: you are brilliant, and the earth is hiring. The earth couldn't afford to send recruiters or limos to your school. It sent you rain, sunsets, ripe cherries, night-blooming jasmine, and that unbelievably cute person you are dating. Take the hint. And here's the deal: forget that this task of planet-saving is not possible in the time required. Don't be put off by

3

people who know what is not possible. Do what needs to be done, and check to see if it was impossible only after you are done.

When asked if I am pessimistic or optimistic about the future, my answer is always the same: if you look at the science about what is happening on earth and aren't pessimistic, you don't understand the data. But if you meet the people who are working to restore this earth and the lives of the poor, and you aren't optimistic, you haven't got a pulse. What I see everywhere in the world are ordinary people willing to confront despair, power, and incalculable odds in order to restore some semblance of grace, justice, and beauty to this world. The poet Adrienne Rich wrote, "So much has been destroyed / I have cast my lot with those / who, age after age, perversely, with no extraordinary power, / reconstitute the world." There could be no better description. Humanity is coalescing. It is reconstituting the world, and the action is taking place in schoolrooms, farms, jungles, villages, campuses, companies, refugee camps, deserts, fisheries, and slums.

You join a multitude of caring people. No one knows how many groups and organizations are working on the most salient issues of our day: climate change, poverty, deforestation, peace, water, hunger, conservation, human rights, and more. This is the largest movement the world has ever seen. Rather than control, it seeks connection. Rather than dominance, it strives to disperse concentrations of power. Like Mercy Corps, it works behind the scenes and gets the job done. Large as it is, no one knows the true size of this movement. It provides hope, support, and meaning to billions of people in the world. Its clout resides in ideas, not in force. It is made up of teachers, children, peasants, businesspeople, rappers, organic farmers, nuns, artists, government workers, fisherfolk, engineers, students, incorrigible writers, weeping Muslims, concerned mothers, poets, doctors without borders, grieving Christians, street musicians, the president of the United States of America, and, as the writer David James Duncan would say, the Creator, the One who loves us all in such a huge way.

There is a rabbinical teaching that says if the world is ending and the Messiah arrives, first plant a tree and then see if the story is true. Inspiration is not garnered from the litanies of what may befall us; it resides in humanity's willingness to restore, redress, reform, rebuild, recover, reimagine, and reconsider. "One day you finally knew / what you had to do, and began, / though the voices around you / kept shouting / their bad advice" is Mary Oliver's description of moving away from the profane toward a deep sense of connectedness to the living world.

Millions of people are working on behalf of strangers, even if the evening news is usually about the death of strangers. This kindness of strangers has religious, even mythic origins, and very specific eighteenth-century roots. Abolitionists were the first people to create a national and global movement to defend the rights of those they did not know. Until that time, no group had filed a grievance except on behalf of itself. The founders of this movement were largely unknown—Granville Sharp, Thomas Clarkson, Josiah Wedgwood—and their goal was ridiculous on the face of it: at that time three out of four people in the world were enslaved. Enslaving each other was what human beings had done for ages. And the abolitionist movement was greeted with incredulity. Conservative spokesmen ridiculed the abolitionists as liberals, progressives, do-gooders, meddlers, and activists. They were told they would ruin the economy and drive England into poverty. But for the first time in history a group of people organized themselves to help people they would never know, from whom they would never receive direct or indirect benefit. And today tens of millions of people do this every day. It is called the world of nonprofits, civil society, schools, social entrepreneurship, nongovernmental organizations, and companies who place social and environmental justice at the top of their strategic goals. The scope and scale of this effort is unparalleled in history.

The living world is not "out there" somewhere but in your heart. What do we know about life? In the words of biologist Janine Benyus, life creates the conditions that are conducive to life. I can think of no better motto for a future economy. We have tens of thousands of abandoned homes without people and tens of thousands of abandoned people without homes. We have failed bankers advising failed regulators on how to save failed assets. We are the only species on the planet without full employment. Brilliant. We have an economy that tells us that it is cheaper to destroy the earth in real time than to renew, restore, and sustain it. You can print money to bail out a bank but you can't print life to bail out a planet. At present we are stealing the future, selling it in the present, and calling it gross domestic product. We can just as easily have an economy that is based on healing the future instead of stealing it. We can either create assets for the future or take the assets of the future. One is called restoration and the other exploitation. And whenever we exploit the earth we exploit people and cause untold suffering. Working for the earth is not a way to get rich—it is a way to *be* rich.

The first living cell came into being nearly four billion years ago, and its direct descendants are in all of our bloodstreams. Literally—you are breathing molecules this very second that were inhaled by Moses, Mother Teresa, and Bono. We are vastly interconnected. Our fates are inseparable. We are here because the dream of every cell is to become two cells. And dreams come true. In each of you are one quadrillion cells, 90 percent of which are not human cells. Your body is a community, and without those other microorganisms you would perish in hours. Each human cell has four hundred billion molecules conducting millions of processes between trillions of atoms. The total cellular activity in one human body is staggering: one septillion actions at any one moment, a one with twenty-four zeros after it. In a millisecond, our body has undergone ten times more processes than there are stars in the universe, which is exactly what Charles Darwin foretold when he said science would discover that each living creature is a "little universe, formed of a host of self-propagating organisms, inconceivably minute and as numerous as the stars of heaven."

So I have two questions for you all: first, can you feel your body? Stop for a moment. Feel your body. One septillion activities going on simultaneously, and your body does this so well you are free to ignore it, and wonder instead when this speech will end. You can feel it. It is called life. This is who you are. Second question: who is in charge of your body? Who is managing those molecules? Hopefully not a political party. Life is creating the conditions that are conducive to life inside you, just as in all of nature. Our innate nature is to create the conditions that are conducive to life. What I want you to imagine is that collectively, humanity is evincing a deep innate wisdom in coming together to heal the wounds and insults of the past.

Ralph Waldo Emerson once asked what we would do if the stars only came out once every thousand years. No one would sleep that night, of course. The world would create new religions overnight. We would be ecstatic, delirious, made rapturous by the glory of God. Instead, the stars come out every night and we watch television.

This extraordinary time when we are globally aware of each other and the multiple dangers that threaten civilization has never happened, not in a thousand years, not in ten thousand years. Each of us is as complex and beautiful as all the stars in the universe. We have done great things and we have gone way off course in terms of honoring creation. You are graduating to the most amazing, stupefying challenge ever bequeathed to any generation. The generations before you failed. They didn't stay up all night. They got distracted and lost sight of the fact that life is a miracle every moment of your existence. Nature beckons you to be on her side. You couldn't ask for a better boss. The most unrealistic person in the world is the cynic, not the dreamer. Hope only makes sense when it doesn't make sense to be hopeful. This is your century. Take it and run as if your life depends on it.

❧

Paul Hawken is a renowned entrepreneur, visionary environmental activist, and author of many books, most recently *Blessed Unrest: How the Largest Movement in the World Came into Being and Why No One Saw It Coming.* He was presented with an honorary doctorate of humane letters by University of Portland president Father Bill Beauchamp, CSC, in May 2009, when he delivered this speech. Our thanks especially to Erica Linson for her help making that moment possible. His Web site is www.paulhawken.com.

Awakening to Our Evolutionary Responsibility

ANDREW COHEN

At the beginning of the twenty-first century, God is no longer "up there" ready to save us. Until very recently, that creative principle was something that *we* would ask for help *from*. But I believe we have reached a time in history when God, which I would describe as the energy and intelligence that initiated the creative process, is now completely dependent upon *us*—upon sentient life forms that have evolved to the point where they are blessed with the extraordinary gifts of self-awareness and freedom of choice. At this critical juncture, our own future and the future of our planet will be determined by the conscious choices that we human beings make, rather than by the whim of a higher power or according to some predestined plan.

The mythical god has fallen out of the sky, and as more and more of us awaken to this fact, it begins to dawn on us that the future is literally in our own hands. Our power and impact have never been greater. It has even been suggested that at this point in evolution, *the process of natural selection has been superseded by human choice.* The decisions that *we* are making, whether we are aware of it or not, have become the primary force directing our planet's future. Indeed, we have become gods. In pre-modern times, gods were revered as supreme beings who had the power to create and destroy life. Who has that power now? We do. We can create life in a laboratory. And we have at our fingertips the power to destroy all life on earth. Our unique capacity to innovate and our drive to create have brought forth unimagined potentials and simultaneously carried us to the very brink of self-destruction. Because our technological development has outpaced our spiritual and moral development, we find ourselves in a crisis of unparalleled proportions. And there is nobody "up there" who can help us now, no supernatural power guiding everything from on high, no God that is separate from our very own consciousness.

For the human being who begins to realize this, the implications on a personal level are profound and overwhelming. Those of us at the leading edge are beginning to awaken to the fact that we are not the people we thought we were. We are not merely individuals with a unique personal history, from a particular family, with a certain ethnic background, who may even recognize ourselves to be global citizens. We are the leading edge of a fourteen-billion-year cosmic process, and the energy and intelligence that initiated that process is now dependent upon us. Our responsibility for being here now and creating the future is much bigger than we could ever have imagined.

Most of us are simply not conscious of the vast evolutionary context in which our own choices and actions are occurring. As long as those of us who are the most fortunate, educated, wealthy, and cognitively developed are lost in the postmodern psychological disease of narcissism, materialism, and self-concern, we will never be in a position to appreciate what the power of choice actually means. But when we awaken to the enormity of the process we are a part of and begin to recognize the extraordinary significance of our freedom to choose, our enlightenment has begun. *There is no other species that has the power to choose to evolve.*

For the mature human being at the beginning of the twenty-first century, this is what I see as being the purpose of spiritual development in a nutshell: to liberate the miraculous power of human choice from unconsciousness and petty self-concern. Our evolution, if not our very survival, depends on it. More and more of us need to awaken to the vast evolutionary context in which we have emerged, and be ready to take responsibility for its overwhelming implications. We must begin to live the gift of human life and use the gift of human choice for the sake of evolution itself.

Andrew Cohen is a spiritual teacher and founder of *EnlightenNext* magazine (formerly *What Is Enlightenment*). A visionary thinker, Cohen is widely recognized for his original contribution to the field of evolutionary spirituality. Through his talks, publications, and dialogues with leading philosophers, mystics, and activists, he has become a defining voice in an international alliance of individuals and organizations committed to the transformation of human consciousness and culture. In 2009, Cohen launched the EnlightenNext Discovery Cycle, an integrated program of spiritual retreats, conferences, and forums for individual and collective evolution. For more information, visit www.andrewcohen.org.

The Original
Human Vocation

BARBARA KINGSOLVER

I can't exactly explain what we're looking for," I told our guests, feeling like a perfectly idiotic guide. "Your eye kind of has to learn for itself."

We were on one of our farm's steepest hillsides, deep in woods, scanning the dry leaf-colored ground for dry leaf-colored mushrooms. My husband Steven found the first patch of morels, a trio tilted at coy angles like garden gnomes. We all stood staring, trying to fix our vision. The color, the shape, the size, everything about a morel resembles a curled leaf lying on the ground among a million of its kind. Even so, the brain perceives, dimly at first and then, after practice, with a weirdly trenchant efficiency. You spot them before you know you've seen them.

This was the original human vocation: finding food on the ground. We're wired for it. It's hard to stop, too. Our friends Joan and Jesse had traveled a long way that day, and their idea of the perfect host might not be a Scoutmaster type who makes you climb all over a slick, pathless mountainside with cat briars ripping your legs. But they didn't complain, even as rain began to spit on our jackets and we climbed through another maze of wild grapevines and mossy logs. "We could go back now," I kept saying. They insisted we keep looking.

After the first half hour we grew quiet, concentrating on the ground, giving each other space for our own finds. It was a rare sort of afternoon. The wood rushes and warblers, normally quiet once the sun gets a good foothold, kept blurting out occasional pieces of song, tricked into a morning mood by the cool, sunless sky. Pileated woodpeckers pitched ideas to one another in their secret talking drum language. These giant, flamboyant woodpeckers are plentiful in our woods. We all took note of their presence and were drawn out of our silence to comment on the remarkable news about their more gigantic first cousins, the ivory-billed woodpeckers. These magnificent creatures, the "Lord God Birds" as they used to be called in the South, had been presumed extinct for half a century. Now a reputable search team had made an unbelievable but well-documented announcement: ivorybills were still alive, deep in a swamp in Arkansas, Lord God.

Was it true? A mistake or a hoax? Was it just one bird, or a few, maybe even enough for the species to survive? These were still open questions, but they were headliner questions, inspiring chat rooms and T-shirts and a whole new tourist industry in swampy Arkansas. People who never gave a hoot about birds before cared about this one. It was a miracle, capturing our hopes. We so want to believe it's possible to come back from our saddest mistakes, and have another chance.

"How do you encourage people to keep their hope," Joan asked, "but not their complacency?" She was deeply involved that spring in producing a film about global climate change, and preoccupied with striking this balance. The truth is so horrific: we are marching ourselves to the maw of our own extinction. An audience that doesn't really get that will amble out of a theater unmoved, go home and change nothing. But an audience that *does* get it may be so terrified they'll feel doomed already. They might walk out looking paler, but still do nothing. How is it possible to inspire an appropriately repentant stance toward a planet that is really, really upset?

I was stumped on the answer to that, as I'd been earlier on the mushroom guidance. However much we despise the monstrous serial killer called global warming, it's hard to bring charges. We cherish our fossil-fuel-driven conveniences, such as the computer I am using to write these words. We can't exactly name-call this problem, or vote it away. The cure involves reaching down into ourselves and pulling out a new kind of person. The practical problem, of course, is how to do that. It's impossible to become a fuel purist, and it seems like a failure to change our ways only halfway, or a pathetic 10 percent. So why even try? When the scope of the problem seems insuperable, isn't it reasonable just to call this one, give it up, and get on with life as we know it?

I do know the answer to that one: that's called child abuse. When my teenager worries that her generation won't be able to fix this problem, I have to admit to her that it won't be up to her generation. It's up to mine. This is a now-or-never kind of project. But a project, nevertheless.

Global-scale alteration from pollution didn't happen when human societies started using a little bit of fossil fuel. It happened after unrestrained growth, irresponsible management, and a cultural refusal to assign any moral value to excessive consumption. Those habits can be reformed. They *have* been reformed: several times in the last century we've learned that some of our favorite things like DDT and the propellants in aerosol cans were rapidly unraveling the structure and substance of our biosphere. We gave them up and reversed the threats. Now the reforms required of us are more systematic, and nobody seems to want to go first. (To be more precise, the USA wants to go last.) Personally, I can't figure out how to give up my computer, but I am trying to get myself into a grid fueled by wind and hydropower instead of strip-mined coal. I could even see sticking some of the new thin-film photovoltaic panels onto our roof, and I'm looking for a few good congressmen or women who'd give us a tax credit for that. In our community and our household we now have options we didn't know about five years ago: hybrid vehicles, geothermal heating. And I refused to believe that a fuel-driven food industry was the only hand that could feed my family. It felt good to be right about that.

I share with almost every adult I know this crazy quilt of optimism and worries, feeling locked into certain habits but keen to change them in a right direction. And the tendency to feel like a jerk for falling short of absolute conversion. I'm not sure why. If a friend had a coronary scare and finally started exercising three days a week, who would hound him about the other four days? It's the worst of bad manners—and self-protection, I think, in a nervously cynical society—to ridicule the small gesture. These earnest efforts might just get us past the train-wreck of the daily news, or the anguish of standing behind a child, looking with her at the road ahead, searching out redemption where we can find it: recycling or carpooling or growing a garden or saving a species or *something*.

Small, stepwise changes in personal habits aren't trivial. Ultimately they will, or won't, add up to having been the thing that mattered.

We all went crazy over finding the ivorybill because he is the Lord God's own redheaded whopper of a second chance. Something can happen for us, it seems, or through us, that will stop this earthly unraveling and start the clock over. Like every creature on earth, we want to make it too. We want more time.

❧

Barbara Kingsolver is the author of seven works of fiction, including the novels *The Lacuna, The Poisonwood Bible, Animal Dreams,* and *The Bean Trees,* as well as books of poetry, essays, and creative nonfiction. Her most recent work of nonfiction is the enormously influential bestseller *Animal, Vegetable, Miracle: A Year of Food Life.* Kingsolver's work has been translated into more than twenty languages and has earned literary awards and a devoted readership at home and abroad. In 2000, she was awarded the National Humanities Medal, the highest honor in the United States for service through the arts. She lives with her family on a farm in southern Appalachia. For more information, visit her Web site at www.animalvegetablemiracle.com.

Living with Losing

You are not going to like what I am about to say, but it's the truth.

We have lost. Those of us who have tried to save the world from the depredations of our own species have lost.

Abbie Hoffman knew this. He put it in his suicide note. "It's too late. We can't win. They've gotten too powerful."

Aye, too powerful. More than that, too many. Just too many people in the world. Thomas Malthus was a rather nasty fellow, I'm told, but he was right:

> *The spirit of benevolence, cherished and invigorated by plenty, is repressed by the chilling breath of want. The hateful passions that had vanished reappear. The mighty law of self-preservation expels all the softer and more exalted emotions of the soul. The temptations to evil are too strong for human nature to resist. The corn is plucked before it is ripe, or hidden away in unfair proportions, and the whole black train of vices that belong to falsehood are immediately generated. Provisions no longer flow in for the support of the mother with a large family. The children are sickly from insufficient food. The rosy flush of health gives place to the pallid cheek and hollow eye of misery. Benevolence, yet lingering in a few bosoms, makes some faint expiring struggles, till at length self-love resumes his wonted empire and lords it triumphant over the world.**

---

* From *An Essay on the Principle of Population,* chapter 10.

This is what we conservationists are up against. This is why we cannot make things better. This is why our well-reasoned letters to politicians, our appeals to the greater public, our appeals to the greater good— this is why it all fails. The human population has built up to the point at which we are behaving like rats in a crowded cage, and nothing is going to improve until our numbers get fewer.

Which they will, one way or another. "Fewer" could very well be "zero." Alas, I'm pretty sure that the required population cut won't occur until we have inflicted far worse on the world than we have thus far.

The environmental movement has been slow to grasp this. Hardly any of my colleagues in the Canadian Parks and Wilderness Society or the Alberta Wilderness Association head their list of must-dos with "Reduce population." They don't seem to realize that even if we were to change our ways and become thoroughly green, at current population levels the world's ecosystems will still collapse. The salient studies are many, but I need cite only one: NASA has found that we now appropriate 20 percent of the earth's annual plant growth to supply ourselves— just one species among many millions—with food, fiber, wood, and fuel.* This cannot last.

We should also recognize that our destructive behavior is built right in. We are naturally habitat-modifiers, like beavers and elephants. We probably cannot do otherwise, dependent as we are on tools and intellect rather than on fangs and claws. Unlike the food-gathering equipment of other species, ours is technological and not self-limiting. It goes out of control as easily as our birth rate does.

Nonetheless, let us assume that through education and legislation and enforcement we could reduce our individual impact significantly. Yet in our billions we could not do so enough to allow normal survival rates for

---

*This is work by Marc Imhoff and Lahouari Bounoua. Read it at www.nasa.gov/vision/earth/environment/0624_hanpp.html. Their figures do not include the oceans. Recent research by Boris Worm, of Dalhousie University, has shown that populations of all marine species we use as food, anywhere in the world, will have collapsed—shrunk to less than 10 percent of historical numbers—by 2048 (*Science,* November 3, 2006).

other species, on which we depend to give us breathable air, drinkable water, and a climate more hospitable than that of Venus.

These arguments are academic anyway. The world's troublemakers are too busy laying waste to think about this kind of thing at all. Reasoning is for university professors, not for presidents ramping up the next war. They will never understand, and they are in charge.

Thus are we stuck. We need to de-crowd the world in order to stop perilous crowded-world behavior, but perilous crowded-world behavior is preventing us from de-crowding the world. I can't think of any way to solve this circular problem, and no one else seems to have figured it out either. The time allotted for answering the question has run out. So I'm going to be honest with you. It looks as if we have had it. The world is already experiencing an "extinction event," as my fellow geologists coldly refer to it, something like the asteroid impact that claimed about two-thirds of the species on the planet sixty-five million years ago. This time we are the asteroid. Our collision with the planet's ecosystems is going to take us down too, and soon. It may happen in one go, via mushroom cloud and other weaponry, or it may be more gradual, requiring a generation or two of proper Malthusian misery as everything goes haywire. Either way, the complex systems required to feed our huge population will fail, and our numbers will crash. Either way, the future looks grave indeed.

The truth may set one free, but this particular truth is pretty hard to handle. It is so crushingly hopeless. So damned sad. Job number one for any organism is to maintain its own kind, yet here we have the entire human race headed over a cliff, and there is nothing a single person can do to stop it. Or even millions of people acting together. Millions more, ignorant and malevolently led, will resist ferociously. This thing has been building up for ten thousand years, through countless wars, tyrannies, insurrections, counter-revolutions, genocides, famines, and plagues ... a long and painful journey from one overpopulation-induced horror to the next. The edge of the precipice looms, our speed is increasing, and the brakes have been disabled by madmen.

How does one deal with that?

I deal with it as most people do when they have to live with wrongs that can't be put right: I choose to ignore it most of the time. Otherwise I would lose my mind, à la Don Quixote, and go tilting at Walmarts.

Instead, I buy things there. Only as second or third choice, you understand, but there it is: the ability to look the other way and carry on. This comes so naturally to us that it must have survival value. Think of everyone at Auschwitz, prisoners and gas-chamber attendants alike, all doing their chores and counting the days until either the Holocaust was complete or the liberators arrived. At the end of the war, there were surprisingly large numbers of both parties still alive.

Knowing what I do about the impending fate of humanity, sometimes I feel like I'm trapped in an upscale extermination camp. Yet I still do my job and pay my taxes, part of the mass of humanity quietly going about its business, ironic proof that we are basically good-hearted and optimistic beings. I live in faint hope that something unexpected and unifying will occur, such that we all wake up one morning knowing that together we can beat this thing.

And I keep writing stuff like this, hoping faintly that it might help to bring on that unexpected and unifying event. Yet every time I write, it puts me in mind of that old Arab saying, "The dogs bark, but the caravan moves on."

Sigh. After blaming myself and my fellow barking dogs for failing to stop the caravan, I now know that doing so is well nigh impossible.

However, this is strangely relieving. The problem of defeatism (giving up too easily) is no longer an issue. We have tried hard and done our best, but we are defeated, plain and simple. Being defeated simplifies things. Strategic thinking—"If we do thus-and-so, maybe we can win!"— is no longer required. More than ever, I can be directed by my conscience. I can now say and do what I believe to be right, even when it doesn't appear to advance my cause, because the cause is lost. How odd: in a personal sense, I have won.

What I have won is a surprisingly good life. My wife and I live in the middle of a national park in the Canadian Rockies. As a self-employed professional naturalist and the author of some popular books on the mountains, I am often hired as a guide by park visitors. This is a lot of fun, especially when I take my clients hiking and backpacking. In the winter months I read to classes of schoolchildren from my novel about ravens. (This is even more fun. We get to make noisy bird sounds in the library.) Much of what I do for a living is enjoyable and appreciated by others. Given the conditions under which so much of humanity suffers, I am lucky beyond words. I have all three things needed to make me happy: I live in a place I love, with people I love, doing work I love to do.

Wishing to sustain this situation leads me to keep sparring with the local despoilers. Unable to really stop them, I try to slow them down as they plunder Jasper National Park, my home, for money. I keep on keeping on, and perhaps you should, too. Why?

First reason: The world is worth it. Our species, remarkable and admirable in so many ways, is worth it. Mostly, though, Mother Nature is worth it. No matter how beleaguered she is, there is always beauty to be found in her. If I can help to preserve little bits of the natural world, those places will provide pleasure to anyone who goes there, including me. And as the extinction event comes on more strongly, protected areas may make all the difference to the survival of species other than our own.

Second reason: An irrational but compelling sense of duty. Thus did the firefighters rush into the flaming towers of the World Trade Center. Thus does the conservationist take on the coal companies. (I did so and lost.) Sometimes the lone good guy wins. And win or lose, good guys inevitably receive awards—sometimes posthumously—for trying.

Third reason: Liberal guilt. It's not fair that my species is wiping out so many other species. It's not their fault that our private party is ruining the planet. And that grieves me. I owe it to the wolverines to give them a chance at survival.

Fourth reason: Wolverines have rights. This is an argument I'm still wrestling with, but if the wolverines ever get lawyers I'd rather be on their good side.

Fifth reason: Encouraging people to protect the environment and have fewer children can't hurt. It's bound to be doing some good, because it's keeping the earth a little greener. The more wildland we can keep intact—and Canada has the most in the world—the better the chance that at least a few human beings will survive the disaster ahead. Perhaps they will be within procreating distance of one another.

Sixth reason: There is always the possibility, remote but still there, that governments may come to their senses and try to turn things around. If so, they will be looking for help. Those of us who have been engaged in eco-related stuff for many years, whether as scientists or activists (or both), have acquired some expertise. We could be useful. In the meantime we can be working on the long list of things that need to be done for planned population reduction to work. These ideas are worth promoting for their own sake, anywhere and everywhere, because they will improve our lives. We can keep pointing to that list every time a politician might be looking.*

Seventh reason: Enjoyment of the game. Taking on the developers can be entertaining. I'm in Canada, where the people across the table are usually polite and do not attack you in the parking lot after the hearing. It's fun to go picketing every now and again, to be on television, and to provide sound bites for the media. Builds poise and self-confidence. Keeps one's protestation skills sharp. And if we don't exercise our right to protest, we will lose it. (Of course, if we do exercise that right in substantial numbers—such that we represent a genuine threat to the established order—then we will lose it, too.)

Eighth reason: The environmental movement has brought some wonderful people through my door. Some have become my friends for life.

Ninth reason: When things get really bad, we eco-buddies can help each other. All those survivalist types squirreling away canned food and guns in their basements are just going to wind up shooting each other. During the worst of times—I'm reminded again of conditions during the great European wars—the key to staying alive has been to surround oneself with trusted family and friends, sharing everything and

---

*The list is too long to include here. It's on my Web site, www.bengadd.com.

looking out for one another. In really dire circumstances, cooperation works better than competition.

Tenth and best reason: Trying to do what's right in this world is a basic human instinct, for most of us a more powerful drive than the temptation to do wrong. Without that built-in altruism, our species would have disappeared long ago. Economist Herman Daly and philosopher John Cobb invented a brilliant new economics based on this finding. It's the subject of their 1989 book *For the Common Good: Redirecting the Economy toward Community, the Environment and a Sustainable Future.* Daly and Cobb disprove the commonly held belief that ending population growth would be economically ruinous. They show just the opposite: that long-term prosperity actually depends on stabilizing our numbers and then reducing them. *For the Common Good* is an important work, right up there with *The Wealth of Nations* and *Das Kapital.* It's also a whole lot more uplifting.

Daly and Cobb agree that lending a hand for the planet's health is its own reward. Trying to keep the land beautiful, the rivers pure, the air sweet—to them that's all just plain good. A no-brainer for anyone, really. Doing right by the earth warms the heart, whether one has much success or not.

That alone would keep me plugging away. But to maintain momentum I have needed one more thing. It's the thing a lot of us Green Party types neglect. It is this: we need to kick back and enjoy the world we are trying to save.

Yes, we need to play, and it helps if it's physical. Too many environmental activists are unfit, urban-dwelling, indoor activists. We need exercise.

We need exercise outdoors, in the natural world. My wife and I are lucky enough to be surrounded by the mountain wilderness we cherish. We can step off our porch and be on the trail in five minutes, enjoying ourselves in a place we have tried hard to protect. After yet another meeting about yet another threat to the national park, when I'm angry with the opportunists gathered at the gates and the park officials who seem much too willing to let them in, there's nothing better than a two-hour hike. It clears the mind and restores the spirit. Evil recedes in the rosy glow of a good workout in natural surroundings.

Why is that? Why is it so attractive to walk in the woods?

I think it's because the wilderness is where our species grew up. That's where we lived back in the days when the world's total population was under a million. Back then we were proud aboriginal hunters and gatherers, not wimpy wage-slaves and Safeway shoppers. We were doing what we had evolved to do, we liked doing it, and the world in which we did it was unspoiled. There were no cities or freeways or coal mines or clearcuts or oil wells or pig farms or car factories or suburbs or strip malls or army bases or missile silos. To quote the Navaho, we "walked in beauty." I think we miss that.

When I'm in the backcountry of Jasper National Park, walking in beauty, the people I meet on the trail might be the same folks with whom I have endured an Edmonton traffic snarl. "Snarl" is right. There, we cursed the situation and each other. In the backcountry, though, walking in beauty, even if it's raining, we smile and say hello. In the wilds we are few, and thus we are nice to each other. It comes naturally and it feels good. The feeling lingers after the trip is over. Great days in the mountains lead to better days back home.

Thus, recreation is an essential part of my life. To make sure I get enough, I have a rule of thirds:

I spend about a third of my time making a living. I have to do that.

I spend another third of my time doing things that I'm not paid for but do anyway, because people I love and care about need the help. This includes everything from household chores to volunteering on worthwhile projects to resisting serious corporate and government misconduct when the need arises. I just can't help doing that.

I spend the remaining third of my time brightening my life, often through physical activity outdoors. I climb the peaks and ride my bike, hike with family and friends, go cross-country skiing and so on. I can allow myself that.

Observing this rule has counterbalanced the doomsday negatives in my life with positives. It helps to ward off the gray waves of despair when they sweep in. If I have learned what is wrong with the world, I am grateful also to have learned what is right. I can live with that.

❧

Ben Gadd, born in 1946, is one of Canada's better-known naturalists and Rockies-area writers. Author of the groundbreaking *Handbook of the Canadian Rockies,* Ben has written eight other books and contributed to several more. He has appeared in many television items and several film documentaries on the Rockies. His Web site is www.bengadd.com.

Humanity's Rite of Passage: Oil as an Adolescent Growth Hormone

ANODEA JUDITH

Sooner or later, we all have to grow up. We've heard this a thousand times, as if it were a fact as certain as death and taxes. Continue to put one foot in front of the other and you will eventually get there, wherever "there" might happen to be. Definitions of maturity vary widely, but most people assume that it will just happen by itself, like ripening fruit, with time as the only necessity.

Humanity is in the throes of an adolescent identity crisis. Birthed from the primal world of nature, billions of years in gestation, we have risen out of Stone Age infancy, crawled across the land in teeming toddlerhood, and fought our way through five thousand years of sibling rivalry, to emerge in the present time as teenagers headed for a collective initiation. If we survive it, a glorious adulthood awaits us, with discoveries and capacities beyond our wildest dreams. But if we fail to transform, many humans and well over a million additional species will die.

No authority figures are going to get us out of this one. Our problems cannot be solved by science, nor politics, nor religion alone. With the Mother Goddess having long been denied, and the Big Daddy in charge dangerously lacking in moral conscience, we are being asked both as individuals and as a culture to grow up and become our own authority, networking and collaborating amongst ourselves to find the best answers. This maturation forms a new organizing principle and is simultaneously psychological, ecological, spiritual, political, economic, scientific, and mythic. With so many fronts on the verge of profound change, navigating this shift can be dizzying. Especially when we are coursing with adolescent growth hormones that have their own agenda.

As a child enters adolescence, he or she experiences a rapid growth spurt, which stops when the adolescent reaches adult size. That growth is brought on by a change in hormones that makes the adolescent, well, a little crazy, as any parent would attest. It is accompanied by intense self-absorption and fueled by consuming everything in sight with little regard for the consequences. Is it any wonder that the media refers to us as "consumers" and encourages us to preen ourselves for some imagined popularity contest? Powerful but reckless, we are struggling with the lure and taboo of repressed libido, and just beginning to form equal relationships with the opposite gender—in the workplace, in politics, in the churches and the home. You could even say we have suicidal tendencies—and the means to carry them out.

Once the hormones settle down and the physical growth stops, emotions settle down as well. Intellectual and spiritual growth take precedence. We become young adults, certainly not without challenges, but life moves forward at a saner pace. The intellectual capacities of the Internet and the spiritual revolution that has yoga and meditation classes propagating across the country show that this revolution is already underway.

Oil has been our adolescent growth hormone, enabling this massive growth spurt to an exploding population and planetary civilization. Oil has generated a worldwide transportation industry that fosters cross-cultural contact and connection. Oil has enabled a worldwide trade economy that brings the latest technologies of Japan to the slums of India, trucks bananas from the jungle, flies cheap toys from China and perfume from France, affecting nearly every economy on the planet. Through the distribution of televisions, computers, and cell phones, oil has made possible—for the first time ever—a global network of communication, wiring up the global brain. Oil is a common ingredient in the products and packaging of most everything we buy, enabling large-scale food production and delivery. Without oil, multi-cultural bodies, such as the United Nations, could not bring together political dignitaries to meet face to face and address global issues.

But just as an adolescent's growth hormone doesn't last forever, our oil-based growth spurt is nearing its end. Not only have we reached our adult size as a population, but the demand for cheap oil has already out-paced its production, and the supply will eventually run out—even if population stabilizes. We simply won't have the fuel necessary to con-tinue the endless consumption that is destroying the environment, the frenzied activity that generates stress, and the irrational compulsion toward imperial conquest that requires an oil industry to support it. With the means now in place for maintaining a global network of intelligence through cyberspace and a growing understanding of our entangled fields of consciousness, our growth will now be more spiritual than material: more about information than products, access rather than ownership, networks rather than markets, service rather than exploitation, personal awakening and health as more valuable than fame and fortune. Instead of using the environment to feed our personal ego-system, we can live in service to our common ecosystem. As the oil hormones settle down, not only will the skies be clearer, but perhaps our sanity will return, and we can begin to enter the beginnings of our future adulthood.

That's the good news.

The bad news is that getting from adolescence to adulthood isn't easy. It usually involves a challenging initiation that contains separation and loss, challenges and ordeals. Like all rites of passage, it asks us to *transform or die.* With a global economy and lifestyle addicted to oil, the machinery that drives our civilization will either transform to a new tech-nology or grind to a halt—and the jury is out as to which will occur first. The oil situation, and the "oiligarchy" that rules it, is clearly coming to a head—like the embarrassing pimples on a teenager's face—revealing the impurities within. When combined with the other environmental crises looming on the horizon, the end of oil will bring an urgent and wrenching shift to all aspects of life as we know it. But the problems we face—deforestation, air pollution, global warming, disappearing top-soil, and the ability to subject distant cultures to the terror of warfare—all are fueled by the use of oil.

Initiation has distinct stages: separation, loss, confinement, challenge, transformation, and rebirth. Our impending crises are evolutionary dri-vers that will drive us through each of these stages into our emerging

HOPE BENEATH OUR FEET

awakening—perhaps the greatest awakening in the history of our kind. In the stage of *separation,* we step back from the cultural trance to awaken to new possibilities. We may choose this voluntarily, or it may be thrust upon us through the collapse of economic and social structures we take for granted, bringing the loss of home through fire, flood, or mortgage crisis or the loss of a job through cutbacks. Our *confinement,* voluntary or not, brings us more time for stillness and contemplation—an awakening of a spiritual dimension and discovery of the world within. Meeting the *challenges* of global hardship will force a higher level of cooperation. Disasters may awaken our hearts, inspiring compassionate service, increased local cooperation, and connection. Our *transformation* will take us from an industrial growth society to a service-oriented society, and from an organizing principle based on the love of power to one based on the power of love. The old organizing principle has been around for five thousand years and is based on a parent-child model. The new one is emerging through increased connection on the Internet, access to information and tools of power and communication, and a hunger for values of compassion and meaning. This transformation can bring about a *rebirth* as an integrated planetary civilization, living at a new level of partnership with each other and our environment, with technologies beyond our dreams. These are all stages that await us as the evolutionary growth hormone peaks and ebbs. This initiation is not the first crisis humanity has faced; it will not be our last. But it is the one facing those of us who are alive today. Our future depends on it.

The context of initiation reveals a silver lining to this wrenching shift. Rites of passage teach the initiate that what is essential cannot be lost, and what emerges is indestructible. One never knows ahead of time what an initiation will bring. Perhaps we will be stripped of what we don't need and discover a more sacred and eternal realm. Can we let go of another pair of shoes from China to have more time to walk barefoot on the grass? Can we imagine the sound of birds without it being drowned out by combustion engines? Can we join with neighbors to tear down fences between our houses and plant gardens, so our food doesn't have to be trucked a thousand miles to the grocery store? Will we have more time to read those books, watch those DVDs, and spend time with our children because it's just too expensive to drive to that meeting or event? Can we finally get in shape riding our bicycles to local jobs that serve local communities, while our minds collaborate in the cyberspace of the global brain?

Members of *Homo sapiens sapiens*—a species capable of knowing itself—have lived for well over a hundred thousand years without the machinery we take for granted today. If we live a hundred thousand more, it will only be with greater intelligence, maturity, and eco-based technologies. Maturity is a greater understanding of ourselves, both our limits and our capabilities.

Evolution is the gods' way of creating more gods. We are approaching godlike powers of creation and destruction on a planetary scale, being equally capable of either. Global warming, genetic manipulation, and nuclear weapons have planetary consequences. Limitation requires that creation and destruction be carefully considered (the essence of wisdom). If we survive the initiation and cultivate that wisdom, perhaps we will become as young gods and goddesses, moving from our adolescence to our adulthood, and from the love of power to a far greater power of love.

❦

Anodea Judith, PhD, is a bestselling author, psychotherapist, and evolutionary activist who travels the circuit teaching paths of awakening that address our current challenges. She is best known for her pioneering work with the chakra system, through her books *Wheels of Life* and *Eastern Body, Western Mind.* This article is excerpted from her double award-winning book *Waking the Global Heart: Humanity's Rite of Passage from the Love of Power to the Power of Love.* For more info on her books and teachings, see www.sacredcenters.com.

Why Bother?

Why bother? That really is the big question facing us as individuals hoping to do something about climate change, and it's not an easy one to answer. I don't know about you, but for me the most upsetting moment in *An Inconvenient Truth* came long after Al Gore scared the hell out of me, constructing an utterly convincing case that the very survival of life on earth as we know it is threatened by climate change. No, the really dark moment came during the closing credits, when we are asked to . . . change our light bulbs. That's when it got really depressing. The immense disproportion between the magnitude of the problem Gore had described and the puniness of what he was asking us to do about it was enough to sink your heart.

But the drop-in-the-bucket issue is not the only problem lurking behind the "why bother" question. Let's say I do bother, big time. I turn my life upside-down, start biking to work, plant a big garden, turn down the thermostat so low I need the Jimmy Carter signature cardigan, forsake the clothes dryer for a laundry line across the yard, trade in the station wagon for a hybrid, get off the beef, go completely local. I could theoretically do all that, but what would be the point when I know full well that halfway around the world there lives my evil twin, some carbon-footprint doppelgänger in Shanghai or Chongqing who has just bought his first car (Chinese car ownership is where ours was back in 1918), is eager to swallow every bite of meat I forswear and who's positively itching to replace every last pound of CO_2 I'm struggling to no longer emit. So what exactly would I have to show for all my trouble?

A sense of personal virtue, you might suggest, somewhat sheepishly. But what good is that when virtue itself is quickly becoming a term of derision? And not just on the editorial pages of *The Wall Street Journal* or on the lips of the former vice president, who famously dismissed energy conservation as a mere "sign of personal virtue." No, even in the pages of The *New York Times* and The *New Yorker,* it seems the epithet "virtuous," when applied to an act of personal environmental responsibility, may be used only ironically. Tell me: how did it come to pass that virtue—a quality that for most of history has generally been deemed, well, a virtue—became a mark of liberal softheadedness? How peculiar, that doing the right thing by the environment—buying the hybrid, eating like a locavore—should now set you up for the Ed Begley, Jr., treatment.

And even if in the face of this derision I decide I am going to bother, there arises the whole vexed question of getting it right. Is eating local or walking to work really going to reduce my carbon footprint? According to one analysis, if walking to work increases your appetite and you consume more meat or milk as a result, walking might actually emit more carbon than driving. A handful of studies recently suggested that in certain cases under certain conditions, produce from places as far away as New Zealand might account for less carbon than comparable domestic products. True, at least one of these studies was co-written by a representative of agribusiness interests in (surprise!) New Zealand, but even so, they make you wonder. If determining the carbon footprint of food is really this complicated, and I've got to consider not only "food miles" but also whether the food came by ship or truck and how lushly the grass grows in New Zealand, then maybe on second thought I'll just buy the imported chops at Costco, at least until the experts get their footprints sorted out.

There are so many stories we can tell ourselves to justify doing nothing, but perhaps the most insidious is that whatever we do manage to do, it will be too little too late. Climate change is upon us, and it has arrived well ahead of schedule. Scientists' projections that seemed dire a decade ago turn out to have been unduly optimistic: the warming and the melting are occurring much faster than the models predicted. Now truly terrifying feedback loops threaten to boost the rate of change exponentially, as the shift from white ice to blue water in the Arctic absorbs more sunlight, and warming soils everywhere become more biologically

HOPE BENEATH OUR FEET

active, causing them to release their vast stores of carbon into the air. Have you looked into the eyes of a climate scientist recently? They look really scared.

So do you still want to talk about planting gardens?

I do.

Whatever we can do as individuals to change the way we live at this suddenly very late date does seem utterly inadequate to the challenge. It's hard to argue with Michael Specter, in a recent *New Yorker* piece on carbon footprints, when he says: "Personal choices, no matter how virtuous, cannot do enough. It will also take laws and money." So it will. Yet it is no less accurate or hardheaded to say that laws and money cannot do enough, either; that it will also take profound changes in the way we live. Why? Because the climate-change crisis is at its very bottom a crisis of lifestyle—of character, even. The Big Problem is nothing more or less than the sum total of countless little everyday choices, most of them made by us (consumer spending represents 70 percent of our economy), and most of the rest of them made in the name of our needs and desires and preferences.

For us to wait for legislation or technology to solve the problem of how we're living our lives suggests that we're not really serious about changing—something our politicians cannot fail to notice. They will not move until we do. Indeed, to look to leaders and experts, to laws and money and grand schemes to save us from our predicament represents precisely the sort of thinking—passive, delegated, dependent on specialists for solutions—that helped get us into this mess in the first place. It's hard to believe that the same sort of thinking could now get us out of it.

Thirty years ago, Wendell Berry, the Kentucky farmer and writer, put forward a blunt analysis of precisely this mentality. He argued that the environmental crisis of the 1970s—an era innocent of climate change; what we would give to have back *that* environmental crisis!—was at its heart a crisis of character and would have to be addressed first at that level: at home, as it were. He was impatient with people who wrote checks to environmental organizations while thoughtlessly squandering fossil fuel in their everyday lives—the 1970s equivalent of people buying carbon offsets to atone for their Tahoes and Durangos. Nothing was likely to change until we healed the "split between what we think and what we do." For Berry, the "why bother" question came down to a moral imperative: "Once our personal connection to what is wrong becomes clear, then we have to choose: we can go on as before, recognizing our dishonesty and living with it the best we can, or we can begin the effort to change the way we think and live."

For Berry, the deep problem standing behind all the other problems of industrial civilization is "specialization," which he regards as the "disease of the modern character." Our society assigns us a tiny number of roles: we're producers (of one thing) at work, consumers of a great many other things the rest of the time, and then once a year or so we vote as citizens. Virtually all of our needs and desires we delegate to specialists of one kind or another—our meals to agribusiness, health to the doctor, education to the teacher, entertainment to the media, care for the environment to the environmentalist, political action to the politician.

As Adam Smith and many others have pointed out, this division of labor has given us many of the blessings of civilization. Specialization is what allows me to sit at a computer thinking about climate change. Yet this same division of labor obscures the lines of connection—and responsibility—linking our everyday acts to their real-world consequences, making it easy for me to overlook the coal-fired power plant that is lighting my screen, or the mountaintop in Kentucky that had to be destroyed to provide the coal to that plant, or the streams running crimson with heavy metals as a result.

Of course, what made this sort of specialization possible in the first place was cheap energy. Cheap fossil fuel allows us to pay distant others to process our food for us, to entertain us, and to (try to) solve our problems, with the result that there is very little we know how to accomplish for ourselves. Think for a moment of all the things you suddenly need to do for yourself when the power goes out—up to and including entertaining yourself. Think, too, about how a power failure causes your neighbors—your community—to suddenly loom so much larger in your life. Cheap energy allowed us to leapfrog community by making it possible to sell our specialty over great distances as well as summon into our lives the specialties of countless distant others.

Here's the point: cheap energy, which gives us climate change, fosters precisely the mentality that makes dealing with climate change in our own lives seem impossibly difficult. Specialists ourselves, we can no longer imagine anyone but an expert, or anything but a new technology or law, solving our problems. Al Gore asks us to change the light bulbs because he probably can't imagine us doing anything much more challenging, like, say, growing some portion of our own food. We can't imagine it, either, which is probably why we prefer to cross our fingers and talk about the promise of ethanol and nuclear power—new liquids and electrons to power the same old cars and houses and lives.

The "cheap-energy mind," as Wendell Berry called it, is the mind that asks "Why bother?" because it is helpless to imagine—much less attempt—a different sort of life, one less divided, less reliant. Since the cheap-energy mind translates everything into money, its proxy, it prefers to put its faith in market-based solutions—carbon taxes and pollution-trading schemes. If we could just get the incentives right, it believes, the economy will properly value everything that matters and nudge our self-interest down the proper channels. The best we can hope for is a greener version of the old invisible hand. Visible hands it has no use for.

But while some such grand scheme may well be necessary, it's doubtful that it will be sufficient or that it will be politically sustainable before we've demonstrated to ourselves that change is possible. Merely to give, to spend, even to vote, is not to do, and there is so much that needs to be done—without further delay. In the judgment of James Hansen, the NASA climate scientist who began sounding the alarm on global warming twenty years ago, we have only ten years left to start cutting—not just slowing—the amount of carbon we're emitting or face a "different planet." Hansen said this more than two years ago; now two years have gone by, and nothing of consequence has been done. So: eight years left to go and a great deal left to do.

Which brings us back to the "why bother" question and how we might better answer it. The reasons not to bother are many and compelling, at least to the cheap-energy mind. But let me offer a few admittedly tentative reasons that we might put on the other side of the scale:

If you do bother, you will set an example for other people. If enough other people bother, each one influencing yet another in a chain reaction of behavioral change, markets for all manner of green products and alternative technologies will prosper and expand. (Just look at the market for hybrid cars.) Consciousness will be raised, perhaps even changed: new moral imperatives and new taboos might take root in the culture. Driving an SUV or eating a 24-ounce steak or illuminating your McMansion like an airport runway at night might come to be regarded as outrages to human conscience. Not having things might become cooler than having them. And those who did change the way they live would acquire the moral standing to demand changes in behavior from others—from other people, other corporations, even other countries.

All of this could, theoretically, happen. What I'm describing (imagining would probably be more accurate) is a process of viral social change, and change of this kind, which is nonlinear, is never something anyone can plan or predict or count on. Who knows, maybe the virus will reach all the way to Chongqing and infect my Chinese evil twin. Or not. Maybe going green will prove a passing fad and will lose steam after a few years, just as it did in the 1980s, when Ronald Reagan took down Jimmy Carter's solar panels from the roof of the White House.

Going personally green is a bet, nothing more or less, though it's one we probably all should make, even if the odds of it paying off aren't great. Sometimes you have to act as if acting will make a difference, even when you can't prove that it will. That, after all, was precisely what happened in Communist Czechoslovakia and Poland, when a handful of individuals like Vaclav Havel and Adam Michnik resolved that they would simply conduct their lives "as if" they lived in a free society. That improbable bet created a tiny space of liberty that, in time, expanded to take in, and then help take down, the whole of the Eastern bloc.

So what would be a comparable bet that the individual might make in the case of the environmental crisis? Havel himself has suggested that people begin to "conduct themselves as if they were to live on this earth

HOPE BENEATH OUR FEET

forever and be answerable for its condition one day." Fair enough, but let me propose a slightly less abstract and daunting wager. The idea is to find one thing to do in your life that doesn't involve spending or voting, that may or may not virally rock the world but is real and particular (as well as symbolic) and that, come what may, will offer its own rewards. Maybe you decide to give up meat, an act that would reduce your carbon footprint by as much as a quarter. Or you could try this: determine to observe the Sabbath. For one day a week, abstain completely from economic activity: no shopping, no driving, no electronics.

But the act I want to talk about is growing some—even just a little—of your own food. Rip out your lawn, if you have one, and if you don't—if you live in a high-rise, or have a yard shrouded in shade—look into getting a plot in a community garden. Measured against the Problem We Face, planting a garden sounds pretty benign, I know, but in fact it's one of the most powerful things an individual can do—to reduce your carbon footprint, sure, but more important, to reduce your sense of dependence and dividedness: to change the cheap-energy mind.

A great many things happen when you plant a vegetable garden, some of them directly related to climate change, others indirect but related nevertheless. Growing food, we forget, comprises the original solar technology: calories produced by means of photosynthesis. Years ago the cheap-energy mind discovered that more food could be produced with less effort by replacing sunlight with fossil-fuel fertilizers and pesticides, with a result that the typical calorie of food energy in your diet now requires about 10 calories of fossil-fuel energy to produce. It's estimated that the way we feed ourselves (or rather, allow ourselves to be fed) accounts for about a fifth of the greenhouse gas for which each of us is responsible.

Yet the sun still shines down on your yard, and photosynthesis still works so abundantly that in a thoughtfully organized vegetable garden (one planted from seed, nourished by compost from the kitchen, and involving not too many drives to the garden center), you can grow the proverbial free lunch—CO_2-free and dollar-free. This is the most local food you can possibly eat (not to mention the freshest, tastiest, and most nutritious), with a carbon footprint so faint that even the New Zealand lamb council dares not challenge it. And while we're counting carbon, consider too your compost pile, which shrinks the heap of garbage your household needs trucked away even as it feeds your vegetables and sequesters carbon in your soil. What else? Well, you will probably notice that you're getting a pretty good workout there in your garden, burning calories without having to get into the car to drive to the gym. (It is one of the absurdities of the modern division of labor that, having replaced physical labor with fossil fuel, we now have to burn even more fossil fuel to keep our unemployed bodies in shape.) Also, by engaging both body and mind, time spent in the garden is time (and energy) subtracted from electronic forms of entertainment.

You begin to see that growing even a little of your own food is, as Wendell Berry pointed out thirty years ago, one of those solutions that, instead of begetting a new set of problems—the way "solutions" like ethanol or nuclear power inevitably do—actually beget other solutions, and not only of the kind that save carbon. Still more valuable are the habits of mind that growing a little of your own food can yield. You quickly learn that you need not be dependent on specialists to provide for yourself—that your body is still good for something and may actually be enlisted in its own support. If the experts are right, if both oil and time are running out, these are skills and habits of mind we're all very soon going to need. We may also need the food. Could gardens provide it? Well, during World War II, victory gardens supplied as much as 40 percent of the produce Americans ate.

But there are sweeter reasons to plant that garden, to bother. At least in this one corner of your yard and life, you will have begun to heal the split between what you think and what you do, to commingle your identities as consumer and producer and citizen. Chances are, your garden will re-engage you with your neighbors, for you will have produce to give away and the need to borrow their tools. You will have reduced the power of the cheap-energy mind by personally overcoming its most debilitating weakness: its helplessness and the fact that it can't do much of anything that doesn't involve division or subtraction. The garden's season-long transit from seed to ripe fruit—will you get a load of that zucchini?!—suggests that the operations of addition and multiplication still obtain, that the abundance of nature is not exhausted. The single greatest lesson the garden teaches is that our relationship to the planet need not be zero-sum, and that as long as the sun still shines and people still can plan and plant, think and do, we can, if we bother to try, find ways to provide for ourselves without diminishing the world.

❧

For the past twenty years, Michael Pollan has been writing books and articles about the places where the human and natural worlds intersect: food, agriculture, gardens, drugs, and architecture. Pollan is the author, most recently, of *In Defense of Food: An Eater's Manifesto.* His previous book, *The Omnivore's Dilemma: A Natural History of Four Meals,* was named one of the ten best books of 2006 by the *New York Times* and the *Washington Post.* Michael Pollan's Web site is at www .michaelpollan.com.

CHAPTER TWO

A Way Forward

Letter from the Future

VICKI ROBIN

Dear Vicki,

Hello from 2030, from your eighty-five-year-old self. We just got the inter-time communications system up and running, and every one of us alive gets to write one free letter to our younger self. There are so many restrictions on what we can say. I can't tell you exactly what is happening. I can't try to "change history." After 2,500 words they cut me off. You can't write me back. Rules! As you can see, the bureaucrats are still with us, but I understand their reasoning. We've made it into a very decent future but had to cross quite a desert to get here. Out of pure love we'd all like to spare you the suffering and change the past, but the GWC (Global Wisdom Council) says that if we eliminate the stripping-away, we might damage the peace we've made with living here.

Even though I can't steer you (as if you would ever let anyone do that!), I *can* shine a light on the choices you are already making—sort of like, "Nudge nudge, hint hint, step there." I can't tell you about the stunning innovations and twists of fate that got us to a fairly decent 2030. I can only talk to you about what you already know.

If you are about to hit delete, thinking this is a hoax, please at least read this quick and dirty key to your future: *less, local,* and *love.* Use less, live locally, and love other people, because they are what see you through.

Hint One: Save—and Make—Energy

You made a good choice in 2008 to do an Airplane Fast and not fly for a year. The irony of flying around the world to lecture people on sustainable living finally got to you. You learned to travel electronically while letting your body stay more still. From that you started to belong where you are and, as you'll see, community is what the future is all about. I think now of the David Waggoner lines: "Stand still. The trees ahead and the bushes beside you are not lost. Wherever you are is called Here." Here, though, is now everywhere as well. The web is humming on levels you can't imagine. Think of the innovations in the last five years—YouTube, wikis, blogs, webcams. Consider Moore's Law (computing power doubles every eighteen months). Add the intuitive capacities you've seen in young children—and yourself. And contemplate what "here" might mean. Thinking globally and loving locally seems very expansive and free. I am personally so moved to know that ancient cultures are finding their roots and rhythms again without busloads of tourists descending every day. It's like watching the desert bloom in the spring. Planes to us seem like jalopies.

While staying home, you'd also do well to follow those impulses to make home more energy-efficient. Hint: don't buy any more lamps for screw-in bulbs; more efficient lighting is coming soon. Hint: just drive your 50-mpg Honda Insight until it dies; you'll be amazed what's next in mob-tech (that's mobility technology). Here I have to bite my tongue. I'll just say that if someone we know invests in some wind farm venture on her island or in a solar installation business, she might be set for life. It used to be "location, location, location." Now it's "local, local, local." By staying home you will see many opportunities to retrofit your home for a post-Peak-Oil future. You'll also find yourself getting political, because shared solutions for energy are better than just putting solar hot water on your roof—as you will anyway.

Hint Two: Grow Food

We've now studied the behavior of our species in transition and have discovered that a spike in "lawns to lunch" (home garden acreage) is a leading indicator of impending resource constraints. The future casts its shadow in the present for those who pay attention, and when people hanker after land and gardening like they used to hanker after opera and travel, you know a shift is coming. Follow all your impulses to grow food, to organize local food systems, to sidle up to neighbors with lawns and suggest that you could find a young farmer who'd love to turn that useless mono-crop of grass into breakfast, lunch, and dinner. Save seeds. Go ahead, if you want, and buy land to grow food, but frankly you have a talent for growing kale and zucchini—and not much else. Become involved with CSAs (community-supported agriculture). Partner with other singles to do a share. You've been thinking about raising chickens. All I'll say is, "Not a bad idea." Or join that goat coop, take that cheesemaking class, and buy up all the used canning jars at the thrift store. Think food. Dream food. Do food. Eat food (but less).

Hint Three: Make Peace with Your Past—and Future

I'm not going to kid you. Some really hard knocks are coming. Some are just as you imagine, others are not. A way of life based on treating finite resources as infinite is ending, and we are *still* living with the shocks and aftershocks of it. We were slow to move on the mandate of 80 percent reduction of carbon by 2050 and are reaping the consequences. Yes, there have been environmental catastrophes, but there have also been "benestrophes"—unexpected accumulations of good fortune. Yes, many have died; some at their own hands, since living within the means of the planet didn't seem like living at all. Be prepared to live through this, knowing that in the larger scheme of things—and nature—it's quite natural for populations to overshoot and collapse. Death itself isn't as tragic as living in fear of death and allowing suspicion and greed to flourish in your mind. Cultivate a calm and caring attitude, even while you rail inside against it all (I can guarantee you'll rail, weep, get mad ... you're human). Making peace *now* with the future means accepting *now* the many losses that will come, so that you won't be in shock and useless. Be like the musicians on the *Titanic*. Create beauty, because those who will die and those who survive both need that. Clearly, since I'm writing, you and others survive—actually, life is grand. Making peace with the future also means that you will roll with the good stuff ahead.

So here are some things you're doing that I'd suggest you keep doing:

Your practice of frugality—getting the maximum pleasure out of every morsel consumed—puts you in a good position to welcome limits as sanity, not deprivation, and to surf the waves of change. Keep teaching your frugality strategies. A lot of people listen to you. Give them something real to chew on.

Realize that there is nothing wrong in your past—it's all useful. Appreciate everything you've done and see what good can come of it. That goes for your relationships, of course—but I also mean (and I can't say too much about it) the whole exuberance of the oil-enabled industrial growth model. Stay open to the good in every technology and every innovation because they may be precursors of the future "light-structures" (in both senses of "light"). Question your assumptions, abandon your Luddite tendencies, and ask about everything, "What's good about you that brought you into being?"

Use your bright mind to see the opportunities in obstacles. Joe Dominguez used to point out to us (you and me . . . funny to talk with you this way) that when there was 25 percent unemployment in the Depression of the 1930s, 75 percent of the people were employed. Evolution tends to favor the braver—those willing to snatch victory from the jaws of defeat. Pay attention to what is being born, even as you tenderly allow all that is passing away to go.

Be an opportunist—but on behalf of your community. The future is friendly to those who shift from "me" to "we." Which brings me to . . .

Hint Four: Treat Everyone Within Fifty Miles Like You Love Them

You will need them as your friends. They are the raw materials of a sane future, if you want to be purely pragmatic. They are also your brain; alone you'll never know enough to survive, but within fifty miles of home is all the intelligence and information you'll need. If you're friendly and generous these neighbors will come to trust you. Of course, friendliness actually takes guts—not the guts it takes to protest (which you will still do for years), but the guts it takes to risk rejection, care first, forgive, apologize, ask before you attack. In other words, loving the ones you're with requires tolerance, acceptance, and letting go of selfishness. I might also point out that among the three million people within fifty miles of you now are probably every friend, lover, dance partner, big thinker, or young person you'll ever need. Go find them. Trade with them. Network with them. Play with them. Help them through hard times. Share meals and homes. Call them to see how their interview or operation went. Ask them to coach you in reaching for your dreams. Even though they aren't "exotic," they're interesting, remarkable, smart, kind, and skilled. Every one a gem.

Pay attention to "co" words. They are the future. Cooperation. Communion. Community. Collaboration. Communication. Your *Conversation Cafes* don't quite fit the word pattern but they are important for people to practice and learn all the other "co" words. Consolation will also be needed.

Do all you can in pairs and teams. Do work parties and cleaning parties and shedding-stuff parties and investing clubs and buying groups and service groups. The era of the Lone Ranger and the Great Hero is passing. Build community. "If you invite them they will come." Alone you are brittle. Together you are supple.

Hint Five: Pack Your Personal Ark

Just as airlines have a baggage weight limit, to cross the great ocean of time and catastrophe into the future you'll need to pack carefully. What of your current life must you have in a future governed by "less, local, and love"? I can't tell you what's coming but I can say this: scenario A is that you muddle through and your daily life doesn't change that much in twenty-five years. The rich get richer and the poor poorer, but life goes on. Scenario B is that catastrophes (and "benestrophes") do come. Your weather does change, the seas do rise, energy shortages do occur, and the dollar isn't what it used to be. Select what you want for either case. If it's A, well, you'll have the things you need and have shed of a lot of excess baggage. If B, you'll have the things you need—and will need them. Here are some categories to consider:

Seeds: heirloom, open pollinated
Books: reference, how-to, and inspirational

Tools: to build things, fix things, make things, study things, kill things (a rifle, butcher knife, and fishing pole), roll things (wheels save your back and feet)

Clothes: warm and durable layers, good shoes, glitter for parties

Furniture: durable, comfortable, multi-purpose

Household: durable. Really useful things with electrical cords are okay (we've never been without that blender), but hand tools will be needed . . . like wire whisks and wooden spoons and good chopping knives.

Health care: stock up on and freeze must-have prescription drugs, buy basic medical books. You'll be surprised at how little those things that you pop in your mouth are still needed. Remember what Norman Cousins said: "85 percent of all illness is self-limiting," and for the rest, I'd say that painkillers and antibiotics are heaven's gift to the creaky.

Beauty: brushes and combs and creams. Keep all those scarves and earrings (and a coupla lipsticks) to feel pretty, which is water for the soul.

Energy: batteries, yes—but everyone should have one back-up solar panel and/or hand-cranked generator for communications technology. Get a solar cooker. Insulate whatever you live in. Install double-pane windows. Use the last hours of ancient sunlight (Thom Hartmann's name for oil) to create a low-energy environment for the future.

You get the drift. Buy and keep what will last. Buy and keep what has multiple uses (like a knife and pot rather than a Cuisinart and electric rice cooker). You're not packing a real Conestoga wagon, so you can keep everything you have now if you want. Remember your old *Your Money or Your Life* idea of enoughness? Not just survival. Not just adequate. Truly rich in everything from basics to luxuries, but nothing in excess. Shed the surplus early and often. Scenarios A and B both favor living lightly.

HOPE BENEATH OUR FEET

Hint Six: Make Yourself Useful

Heads up. A local future belongs to the person who makes herself truly useful to real people, not to the one who can market some useless gadget to unsuspecting consumers. You'll find it hard to trade your knack for inspiring others for bicycle repair, but don't worry. If you can make people laugh, you'll always be taken care of. Hone all people skills (see Hint Four above). The future needs facilitators, negotiators, re-framers, therapists, counselors—anyone with patience in the face of human suffering. The future also needs handymen, emergency management specialists, nurses, gardeners, inventors, record keepers, geeks and techies of every ilk, musicians, athletes, mechanics, engineers, cooks, team players, canning experts, teachers, midwives, writers, body workers, artists, project managers, story tellers, hunters and fishermen, builders, designers of every sort imaginable, healers of every sort imaginable. There's no lack of good work here in the future.

I do hope this all gets through. The censors may zap anything I say that gives you too much information. But here's what I *can* tell you about now. The birds are singing. The children are healthy. They don't blame us for our mistakes—we now know for certain that our generation did our best with what we had and what we knew. This new generation understands that blame is toxic and they simply don't do it. It makes them seem like angels, really. They know they are making the future—and that's what gives meaning to life. They are actually watching over you now. Yes, we in the future travel in time to care for you. We do our best to help without interfering. You are loved. All of you. Have courage. Keep going. It's working out.

Vicki

ᘉ

Vicki Robin is co-author with Joe Dominguez of the international bestseller *Your Money or Your Life: Transforming Your Relationship with Money and Achieving Financial Independence* (Viking Penguin, 1992), available now in ten languages, and recently updated for the twenty-first century (Penguin Putnam, 2008). Vicki is also an active social change agent. She is co-founder of the New Road Map Foundation, the Center for a New American Dream, Sustainable Seattle, Conversation Cafes, the Simplicity Forum, the Turning Tide Coalition, Let's Talk America, and currently Transition Whidbey. She serves on the board of Transition U.S.

A Way Forward
in an Uncertain Future

SUSAN FEATHERS

It has been nearly three decades since James Lovelock published a scientific premise for the earth as a living organism. He based his work on observations that show a self-regulating zone of life called the "biosphere." *Gaia: A New Way of Viewing the Earth* suggested that living communities regulate the life functions of a living planet.

Human beings are mostly unaware of how temperature, recycling, energy flow, and population are managed on earth, yet humans affect outcomes and are affected in turn by them.

Consider what Lovelock observed about the differences in atmospheric composition among "dead" planets and a "living planet." Planets with high concentrations of carbon dioxide in their atmosphere (Venus and Mars) are either broiling or ice blocks. They contain little or no life. Earth's surface temperature before the advent of life averaged about three hundred degrees Celsius. With the current envelope of life on Earth, the average temperature is thirteen degrees Celsius.

Organisms evolved in oceans that began to remove carbon dioxide from the air and give off oxygen. Gradually nitrogen and oxygen replaced carbon dioxide in the atmosphere. A hydrologic cycle formed, and Earth began to cool to conditions that allowed life as we know it to proliferate.

For nearly two billion years the plankton of the oceans, lakes, and later trees and land plants of Earth sequestered carbon from the air, incorporating it into their forms. As life evolved into more complex forms, animals—passing carbon they ingest from plants or plant eaters consumed—deposited more carbon into "sinks" as their bodies and waste were incorporated into the earth's crust. These accumulations of life forms, whether sinking to the bottom of the sea or dissolving into the crust of land over time, were gradually transformed into rich, black strata: oil and coal.

For a long, long time this was true. Life followed five principles of self-regulation:

- Use of a non-polluting, unlimited energy source;
- Recycling of matter through food webs;
- Preservation of biodiversity in genes, kinds of creatures, habitats;
- Fine control of populations to stay within carrying capacity;
- Change in response to new conditions (evolution).

Then came the conscious being, the human, whose brain began to question how things worked and whose ingenuity mimicked nature. A certain kind of wisdom grew as men and women observed nature's ways and lived accordingly.

Things began to happen in one or two places in the world as man's knowledge grew. Man wondered if combusting oil or coal could help us get things done. And as the pundits say, the rest is history.

A relatively small percent of the total human population has been putting that store of carbon back into the atmosphere as carbon dioxide emissions, and the earth's biosphere is heating up as a consequence.

For readers aware of the dynamic behind the human carbon footprint, it is very frustrating to live beside so many fellow countrymen and women seemingly unconcerned. Recently a woman I met stated that the warming of the earth's surface was "going to happen anyway, and besides, man as a species is not going to be here forever."

Statements like that tell me how little people grasp the essential reality of life: we are not here alone, nor did we arrive here alone, but we evolved in tandem with thousands of species whose lives are intricately connected to ours and that make it possible for us to have life and to have it abundantly.

Americans in particular have never been more disconnected from their biological inheritance. We talk about "nature" in the abstract even as millions of microbes cleanse our skin, digest our food, and destroy harmful invaders. And this is happening in our very own bodies! Microbes "get no respect" in America.

It's not that we are bad people who have set out to plunder the earth, but we are ignorant of how life works and how we are a part of it. That knowledge, once the inheritance of every young child growing up in communities across the earth, has been lost in modern technological cultures—lost to our peril.

So, what can thoughtful people do?

On a personal level we can contemplate those five principles of ecosystems and use them as a checklist for our own lives. How can we use less of a polluting type of energy, recycle more, leave a smaller footprint, create habitat in our yards, join conservation efforts, and change our ways to meet the new challenges? Call it a program of self-regulation.

Some might challenge that suggestion, asking why they should give up their comforts or restrict their activities when no one else seems to be doing so. That reminds me of Albert Schweitzer's quest to find an ethical basis for living. Here is what he thought:

> *As I sit here under this tree I think about how much I value my own life and wish to go on living and to have more of it. Then I look up at this lofty tree with its gently swaying leaves and think, "This tree must hold its own life as valuable and also want to go on living and have more of life, too. And even though it is mute, it nevertheless is no different than me in its desire to live, to grow, to flourish."*

Everywhere we see this, if we stop to observe ... the force of life willing itself into being and survival upon the face of the earth for its time. The fleeing gazelle with the swift cheetah in pursuit, the child bubbling with excitement about being alive, and bees pollinating the flowering beings that bring us pleasure and food—all to survive, to thrive.

Recognizing this common bond to all of life around us, Schweitzer realized, results in Reverence for Life, which he concluded is the ethical basis for living. We begin to value our own life more, to see it as a precious gift and to live it to its highest purpose. We regain the will to live.

This brings us full circle to Lovelock's premise that the earth itself is a living organism of which we are all functional parts. All together the whole thing works. Works, that is, as long as we follow the five basic principles that are the great roots of life on this planet.

Perhaps this time in human history is a call to return to a higher purpose in life, to realize the human's role to *consciously participate* in the well-functioning of earth's living systems.

We are being reminded of our place in the whole pageant of life we find around us. We are called to Reverence for Life as a *way of life.* As individuals we can find emotional, spiritual, and practical guidance from the life we observe around us. Reconnecting, experiencing life in all its manifest forms as fellow inhabitants with which we share this beautiful planet—this life—is a way forward in an uncertain future.

REFERENCES

Lovelock, James. 1995. *The Ages of Gaia: A Biography of Our Living Planet.* New York: W.W. Norton & Co.

Schweitzer, Albert. 1990. *Out of My Life and Thought: An Autobiography.* New York: Henry Holt & Co.

❧

Susan Feathers is an environmental educator and nonprofit consultant whose writing and fundraising for nonprofits is aimed at realizing a nonviolent, ecologically sustainable world community. Visit her Web site for more information: www.writeforchange.com.

Living with Purpose
in the End Times

JAMIE McHUGH

I am by nature an optimistic person. Yet, with the bleak statistics about our environmental crisis, I have lost faith in a positive outcome for the planet. In fact, I have become convinced that we are living in the end times. But this conclusion does not depress me as much as it makes me more committed to living now. And, as I know from my own life story, even statistics are not absolute predictions. Twenty-one years ago, I was diagnosed with HIV disease. The diagnosis was considered a death sentence then, making me confront some challenging questions: How should I live if tomorrow is not guaranteed? If there is no treatment, how do I treat myself?

Humanity has been given a diagnosis—environmental catastrophe—and is facing similar questions. There is no cure on the horizon. So how do we live from day to day? What can we do individually and collectively to treat the illness? Human beings seem to be wired more for rapid response than for sustained action over time. A cancer diagnosis, like a tsunami, is an emergency and requires immediate action. Environmental disease, like HIV, is a gradual process of immune dissolution. It's easy to become complacent and not feel the urgency on a daily basis. In fact, many are not even aware they have been diagnosed. Living with a sense of separateness, they don't take the fate of the planet personally, even though they share the same lifeblood—the air, the water, and the land.

When diagnosed, I had to wake up and take action on my own behalf, as the experts had no viable solutions at the time. I learned to truly love myself and make my health and well-being a priority. I became less inclined to project ahead into the future, and more willing to live a year at a time. I grew content living with less; having more open space for being in tune with my body's rhythms and my creative expression became a greater need than having material goods. Gradually, I found myself living more out of a love for life rather than from a fear of dying.

My friend and mentor George Solomon (one of the pioneers of PNI: psychoneuroimmunology) compiled the research in 1993 on long-term survivors of AIDS. This was prior to the breakthrough drugs, when there were no medical solutions for HIV. The four enduring themes he identified were:

- Healthy Self-care
  (I am responsible for my health and well-being)
- Sense of Perspective
  (I am going to die eventually, but not tomorrow)
- Sense of Purpose
  (I am here on this planet to do something of value)
- Healthy Relationships and Social Connectedness
  (I belong and am loved)

In looking back at these four criteria for health, I see how they can serve us as we all come to terms with our collective diagnosis, as well as the inevitable fear and hopelessness that arise with a terminal illness.

Healthy Self-Care: This is a matter of making choices that are within the realm of my own agency. When I am in the city, I ride my bicycle to get aerobic activity (I live on top of a hill), and driving less is beneficial for the ecosystem of the city. I cook at home and eat well; all of my vegetable scraps are composted and feed my garden. Eating organically is just as much for the health of the land and the water as it is for my body. I recycle and re-use as much as I can to conserve money and materials. My bodily practices of breath, sound, movement, contact, and stillness are sustainable tools for tending to my inner ecosystem and conserving my own energy. My trade-off is using my car to drive every other week to my rural retreat, about three hours away. Being in nature, being close to the earth, shifts my self-perception.

Sense of Perspective: When I am in the wilds, I feel both my insignificance and my vastness. I stand in awe of the power of the planet and am also aware: this is my body! This perspective expands my self-definition and broadens my circle of awareness. An ongoing practice is to go to the ocean at the end of each day for a simple prayer. With arms outstretched as the sun slides into the sea, I say aloud, "Thank you for this day." This spoken declaration affirms my gratefulness for being alive on this planet now. This becomes especially meaningful when it has been a difficult or dispiriting day. Expressing this statement daily is a consistent element in a ritual of renewing my contract with the natural world and forces unseen. Through this practice I am reminded of my own nature, my own place, in the mystery of life. Sensing larger forces and a greater intelligence at work reassures me that life on the planet will prevail in spite of our shenanigans.

Sense of Purpose: Two years ago, I was leading a retreat at the ocean. During a morning meditation, I heard a clear inner voice: "You need to go and live in nature now. You can't wait until you are partnered, or have a plan; just do it now!" So, off I went, bringing along a new digital camera, and found myself re-discovering my creativity through the focus of a lens. I have been photographing for the past thirty years, making art as a personal meditation. One thing led to the next, though, and within the year, I was commissioned to do a shoot, presented my first exhibit in San Francisco, and had three more shows lined up for the following year. My passion for artistic expression and for the planet intersected into a newfound purpose: being an emissary for nature.

Healthy Relationships and Social Connectedness: Integral to a sense of purpose is my place in culture as an artist and teacher, and the connection it provides me with a wide range of people. For years I have led retreats at Point Reyes National Seashore to support people in finding their own relationship to the natural environment through movement and the expressive arts. Creatively exploring the body awakens the senses to the body outside our skin. Doom-and-gloom statistics seem to overwhelm people and increase anxiety and despair. A direct encounter with the natural world, on the other hand, reawakens love and appreciation for the self and the planet. As a long-time environmental educator shared with me at the end of a retreat, "I have been teaching the science of the environment for so long I had forgotten how deeply nature moves my soul. I want to use the arts and senses more from now on with the kids."

Nature is such powerful medicine, and I am moved to share this healing with others. As much as I enjoy taking groups out of the city into nature, I can reach more people by taking my large-format photographs into urban settings, particularly into hospitals and other therapeutic environments, where the soul of nature is needed.

Can we save the planet, and our species, in time? I am not sure, but this much I do know and trust: just as I had to find my own healing solutions and not depend upon experts when first diagnosed, I also need to live my own solutions now and continue to mindfully treat the earth—materially and spiritually.

What good does it do if I eat organically and give money to environmental organizations if I am simultaneously complicit in fouling the air and water through the excesses of my lifestyle? Self-love connects me to my body, and love for the planet connects me to my larger body. The two bodies are really one and the same. If I live mindfully and remember my larger body, I am able to trust life more. If I forget the larger body, I am constrained by the limits of my individual conditioning and more likely to fear the future. Walt Whitman said, "I become multitudes!" As the diagnosis taught me many years ago, survival depends upon participation and communion rather than withdrawal and separation. Creative expression in nature, involvement with the local community, and my daily ritual are all powerful medicine and reminders: I am—and we are—all a necessary part of the unfolding Creation Story.

≫

Jamie McHugh is a registered Somatic Movement Therapist, a performance artist and an award-winning fine art photographer living on the northern California coast. A master teacher of somatics and expressive arts, he has developed "Somatic Expression," teaching body-based work internationally for thirty years to people of all ages. Jamie is currently adjunct faculty in the Holistic Health Department at John F. Kennedy University and at Tamalpa Institute in the San Francisco Bay Area. Visit him at www.somaticexpression.com and www.naturebeing art.org.

Love the Things
We Love

LARRY SANTOYO

To prepare for the future, we can love the present!

Acting and reacting from a place of fear of the future will only lead to the worst of our collective mistakes of the past. I am not being naïve about earth changes and oil—I just don't see the point of fearing the future, because you can never make good decisions based on fear.

Right now I am a permaculture teacher and designer; before that I was one of those hippie back-to-the-landers, and before that I was a cop—yes, an officer of the law. In my years as a peace officer, I was trained to respond to all types of stressful situations with a certain protocol that included calm, clear thinking. Those people who were in distress, who I was trying to help (or subdue), usually were not thinking clearly, could not help themselves and were, in fact, a danger to themselves and to the proverbial others.

I only had the upper hand in any of those cases when I was thinking clearly, calmly, and observing the patterns around us. I was almost always injured or lost control when I reacted out of fear or anxiety.

My experience is that when you are reacting from fear your adrenaline is pumping, you lose peripheral vision, and you can't keep up with individual thoughts as separate signals—so you blur them, and your responses become blurred too, into general sweeps of thought, speech, and action. Some of us default to victim and maybe react or don't react (as the case may be) like "a deer in the headlights." Some of us might overreact—responding with too much information, aggression, or violence. I think that anxiety affects us in the same ways as fear.

The spread of the new eco-anxiety disorder has as much to do with the environment as acrophobia has to do with architecture or geology.

It is simply an example of how some of us react under stress. Uncertainty, change, and even anxiety are stressors in the human environment that will shape us, but we can choose what form we shall take. Practice slowing down—for example, time in the garden or quiet observation in the wilderness—and meditate on the things that make you happy. Do whatever you can to learn this skill set well.

If we choose our actions based on what we love about the present, we then have a solid foundation upon which to build our coming lives. What do we want to see in the world? This should be the question we ask ourselves, not what do we *not* want to see. . . . Do we love eating fruits and vegetables grown without toxic chemicals? Then let's work toward creating the conditions for organic crops to flourish. Do we love waking up to the song of birds in the morning? Then yes, let us create the conditions for that to happen too.

There is actually very little reasonable evidence (without a premise of fallacious speculation) that total catastrophe is coming. Change is coming for sure, but the casualties, destruction, and despair we read about is the doomerism of old. I also see very little evidence that the world will be abruptly brought to a doomerist halt—a bigger bang is not in the equation. . . .

Yes, change is coming, but then, change is always coming. People/culture/humanity have clearly shown the capacity for resilience, for adaptability, and for enduring ingenuity. I have an immutable trust in our abilities as humans for compassion and am grateful for my experiences over the years, where I have seen anonymous acts of true heroism that made me (a grown man) cry. I believe that most people are just waiting to be called to rise to their highest abilities, and that the trumpet call of that challenge is just now reaching their ears.

Living lightly, going green or sustainable, doesn't actually mean giving up most things—it just means changing the way we get them, what they are made of, and by whom. As we focus on the realities of energy "use" itself, we quickly arrive at the conclusion that local is more economical and more sustainable. Thus I agree that one of the first steps toward a sustainable future is localizing, not just food but everything.

"Localize Enterprise" is my mantra now—*all* enterprise. A wandering jack-of-all-trades or handyman/woman will become the most welcomed stranger in the future. Learning a basic vocational trade should be an important objective for all, while cheap energy still remains to allow information to flow so freely and easily.

After working with various communities around the world, I find that the issues of information, transportation, livelihood, food, and water are all identified as fundamental and key, and when solved are core to the function of strong and stable communities.

Not only can communities share common issues, they can also share common solution strategies. I am reminded that if we've been able to solve community survival issues in the remote and adverse conditions of wars and droughts, we can certainly solve them here in the U.S. where we have access to such a wide array of the comforts of information, technology, and the world of multiple choice.

Growing food is not the weak link in the food chain. In the eighteenth century, one or two farmer/horticulturists could provide food for fifty to a hundred hungry villagers—all without petroleum-powered machines or chemical fertilizer. With a swift learning curve, we can do the same.

It's everything besides food production that will be much harder to get and/or get done, if and when there are fewer resources to get 'em done with.... And it will be even harder to get by and/or get it done in the "country." Yup, contrary to popular belief, there are far more options for getting closer to sustainability in urban areas than back on the land....

Most of our urban areas are in proximity to historical settlement sites—where we did live and flourish before the advent of cheap oil. The hidden depth of human ingenuity is almost impossible to comprehend. With today's collection of information and access to historical review of successful land-management strategies, we can develop the practices and redevelop the requisite skill sets to solve core issues that will allow our communities to function again as ecosystems. We can learn to make our decisions not out of fear of the future, but because we love the things we love about the present.

HOPE BENEATH OUR FEET

Larry Santoyo is Vice President of the Permaculture Institute (USA), Co-director of The Terra Foundation, and Director of EarthFlow Design Works, online at www.earthflow.com.

Gandhi Then
and Now

MICHAEL N. NAGLER

All crises are opportunities—if you know how to find them. The crisis we are facing now is huge, simply unprecedented in all of human history; and I believe, logically enough, that the opportunity hidden in its coils is just as great. While there is no *guarantee* that we'll take advantage of this unprecedented opportunity, or even that we will make any changes at all in time to save ourselves, it is perfectly possible that in the face of such peril the world can come to its senses, become aware of itself in a much deeper way than ever before and set human progress on a new course.

We have gained some new understandings only recently of what is really causing this crisis, and why progressive forces have been so unable to affect it. First of all, it is fundamentally a *spiritual* crisis. All the misguided policy we're seeing is embedded in a *culture;* and underlying that culture itself is a disastrously materialistic vision of the human being and reality in general, leading inevitably to emptiness, despair, greed, and violence. I will venture a word about how to address this at the end of this brief chapter, but first, just this year (2007) two aspects of the crisis have become clear that have a direct bearing on strategy:

Negative tactics are not working. "Scare" tactics don't work because fear is part of the problem, not the solution. Appeals to "use less" because of the looming scarcities fall on deaf ears: Americans are simply in no mood to do without, no matter how clearly resources of whatever kind are dwindling. Whatever may have been our culture in centuries past, the barrage of commercial messages to which North Americans are exposed (according to one study, three thousand of them a day) has rendered us, as a people, incapable of renouncing, downsizing, doing with less. We are "more" people; asking us to have less sounds foreign and defeatist.

On the other hand, we are in the midst of an enormous social revolution. As Paul Hawken has found, millions of *organizations,* not to mention people, are doing good work of one kind or another worldwide to rectify the looming disaster: ecological, economic, and conflictual. They are experimenting with local currencies, agitating against wars, buying into community-supported agriculture (CSA), learning alternative medicine, living communally, going off to offer humanitarian aid, learning something about nonviolence—the myriad projects acknowledged each quarter in *YES!* magazine and elsewhere.

However, all this *activity,* laudable as it is, is not—or not yet, anyway—a *movement.* It does not have a common sense of its overriding goal (we know what we're against more than what we're for), an agreed upon strategic plan, or—the importance of this should not be overlooked—a name.

In the light of these features we need to stop asking our fellow Americans (or whomever) to make do with less. Instead we should be offering them the prospect of *shifting their aspirations* to something higher, something that will in fact make them happier, more fulfilled, much more efficiently than the consumption of unnecessary goods or services that brings in its wake the inevitable alienation from and competition with others that we see around us.

Martin Luther King had it exactly right:

"We must change from being a thing-oriented civilization to a person-oriented civilization."

As one of the ancient Upanishads states very simply, "Man [the human being] can never be satisfied by wealth." A variety of studies are now proving this on the scientific level, and many people have reached the point of satiation where they are dimly aware of this truth though there is nothing, no support from the surrounding culture to help them acknowledge, much less deal with it. On the other hand, people are turning to voluntarism and service in unprecedented numbers.

After the disastrous tsunami in Southeast Asia, a United States Marine who had been handing out relief supplies throughout the day was asked how he felt about that kind of work, and he said, "I have been serving my country for thirty-four years and never got any fulfillment until today." What kind of civilization have we built, that makes us spend thirty-four years going in the wrong direction and (if we're lucky) one day of doing something useful? King was exactly right: fulfillment comes from ever-deeper relationships, not from things. And most of us are actually aware of that; in survey after survey the number one thing people say they are longing for is community. This became clear to me in an almost visceral way in just a few weeks that I was able to spend recently in India. Despite the wreckage of the physical environment there, you never feel that *people* are isolated from each other as they are here. Even a very rich person walking past a beggar does not shrink away; even the beggar, if he or she has failed to get a handout, does not glare at him or her in anger. Everywhere people feel that they are part of each other.

I propose that we dare people to rebel against the advertisers and entertainers who are leading us down this empty road of materialism; that we harness American rebelliousness and individuality precisely to overturn the predominant culture in favor of one based on humanity.

Gandhi, after all, built his whole scheme for the world order on this central value: the individual serves the family, the family serves the village, the village serves the district, the district the state, the state the nation, and—important not to leave this step out—the nation the world community. Life will not be a pyramid, he proposed, but "an oceanic circle whose center will be the individual."*

To realize the so-far inchoate movement that all of our good projects have not become, we can configure them, each of us, in such a way that they resonate with and enhance this basic goal of looking for fulfillment in people rather than things, service rather than domination. I like David Solnit's language (I usually like his language): "If we, as a movement of

---

* *The Harijan.* July 28, 1946,. p 236.

movements, adopt a people-power strategic framework ... we will have a viable ... strategy. It's clear that we're not all going to agree on any one (or two or three) campaigns, but it is possible for us to *consciously* adopt ... a strategy that makes our various efforts complementary and cumulative."*

It is not yet clear how we will, for starters, convince people to change their desires from things to relationships with others, but we do know that human beings imitate one another intensely, and that someone who is perceived to be happy is often, though unconsciously, imitated by someone who secretly knows he or she is not. It may not take persuasion so much as *demonstration*. After all, that's how advertising works. We can use that psychological vulnerability, for once, to our advantage.

As Solnit points out, this core "revolution of values" (again King's term) is the foundation, but it is one on which we have to build. We need some strategic overview, and coherence in our behavior and our message. It's not at all clear—to me, at least—how to make this happen, but it's very clear that it *can* happen, and probably with or without a single prominent leader. In these times, it may have to be without such a person: so be it. But we know enough about "tipping points" and "paradigm shifts" today to know that a compelling idea can pull millions of people together, though their participation in carrying it forward may be very diverse.

No matter where you start in thinking about this huge challenge, you realize that we are not alone: Gandhi has been there. He is the one person in the modern world who worked this all out on every level, strategic and principled, from the minutest detail to the grandest overview. Without his inspiration and his guidance I do not believe we can carry off this revolution at all. Let me close with an example.

Gandhi's campaign for India's freedom (and the downfall of colonialism as we knew it) was a balance of *constructive program,* where you go ahead and build the future without waiting for permission from

---

* "People Power: It's Time to Stop the War Ourselves," *YES! Magazine.* Winter 2008,. p 15. Italics added.

your oppressors, and the complementary effort I call *obstructive program,* civil disobedience, or Satyagraha, to resist oppression at key points and propitious times. Ideally the constructive part would do its work so well that the confrontational part would be minimal. (A similar balance is proposed today by Joanna Macy). Gandhi found that constructive program needed a single project that symbolized and gave coherence to the whole, something everyone could take part in, and for his situation the ideal candidate was *khadi:* the spinning, weaving, and distribution of homespun cloth. There were many reasons for the choice, which I go into in my book, *The Search for a Nonviolent Future,* but what gave khadi so much power was that it was non-confrontational but subversive: spinning your own cloth was your own perfectly legal business, but it put British manufactures, and the exploitive system built on them, out of business. Khadi also dealt with a basic need, the second item in "food, clothing, and shelter." And of course it could be locally, indeed individually organized. What would be the equivalent today? After talking this over for years with many audiences, I believe we have come up with a darn good candidate: local food. It's basic, legal (which you sometimes want), local, healthy—and would put one of the most damaging forms of industrialization, agribusiness, out of business.

What, in the end, should every one of us do, along with getting involved in growing or at least consuming local foodstuffs? Here's the list I propose:

- Study Gandhi.
- Opt out of the commercial system. Just don't buy it, either the products or—more importantly—the ideology that products can make us happy. (In other words, ignore the mass media!) Instead,
- Cultivate many and deep relationships, and finally,
- Develop ourselves spiritually. Religion may be a mass phenomenon, but spirituality is not. It is an individual matter—hence its strength—and I will say no more about it here except to encourage you to find, if you have not already done so, a teacher and a practice. It's the most revolutionary act of all.

HOPE BENEATH OUR FEET

Michael N. Nagler is professor emeritus of Classics and Comparative Literature at UC Berkeley, where he founded the Peace and Conflict Studies Program. He is founder and president of the Metta Center for Nonviolence Education (www.mettacenter.org).

Inspiring and Sustaining
Action Over Time

RUSKIN K. HARTLEY

Locked in the rings of an ancient redwood in the depths of Northern California's Humboldt Redwoods State Park lies an unbroken climatic record stretching back 1,800 years to the time of the Roman geographer, Ptolemy. For the past thirty-eight years, I have been part of this record. Imperceptibly, this tree has been recording the imprint I, and my fellow six billion humans, have bequeathed to the atmosphere. By measuring minute changes in the ratio of carbon and oxygen isotopes in tree rings, scientists can reconstruct past climates and begin understanding how this tree has responded. For at least the past hundred years, its rings have borne silent witness to the earth's steadily increasing levels of carbon dioxide. But this tree also bears witness to the power of people to protect nature when mobilized.

The tree's biggest threat arrived less than one human lifespan ago when it was slated for harvest. The fact that it stands strong today is testament to the power of the redwoods to inspire action, and the ability of individuals and communities to mobilize. Rather than being cut and split for grape stakes, this ancient giant was protected in a state park. It continues to be protected today because of a complex web of individual and community actions that are sustained to this day.

While this individual ancient tree will likely survive even rapid climate change, the threat to the coast redwood forest as a whole has never been greater. In one study, scientists reported that coastal fog provided up to 45 percent of a redwood tree's annual water needs. Unfortunately, over the twentieth century it is projected that the incidence of fog on the coast has decreased by perhaps as much as 50 percent. You don't need to be a scientist to understand that if this trend continues these forests that have flourished for millennia risk extinction through water stress.

So what lessons do I take away from this tree to help me navigate the looming environmental catastrophe? The story of the redwoods is that fear may provoke action, but respect and love must sustain that action over time. The tree was initially "saved" when threatened by loggers. But it's been sustained in the state park because it continues to inspire successive generations who value its intrinsic worth above its use as a redwood deck—the twenty-first century equivalent of the early twentieth century grape stake. In addition, while the genesis of its protection was an individual act, lasting protection was ultimately achieved through a combination of extraordinary individual generosity, a community of people organized through Save the Redwoods League, and finally the citizens of the state of California who voted to tax themselves to save the tree and its surrounding forest. Over the last ninety years, this model of the individual, the organization, and the public pulling together has established a system of redwood parks and reserves that span California and protect some of the most glorious forests on earth.

In the face of climate change, a similar mobilization is required to save these forests all over again. For the redwoods to survive, we must ensure that parks are large enough to sustain natural processes, that they are connected to allow for the migration of plant species in response to climate change, and that climate change itself is addressed so it does not breach ecological thresholds beyond which recovery is impossible. This challenge is an order of magnitude more complex than the challenge of sparing the tree in the first place, but the same principles apply: that people will be inspired to save what they love; that meeting this challenge will require individual leadership, community organizing, and a broad public effort organized through our governments; and that this effort must be sustained over many lifetimes to be successful.

While not everyone has a strong connection to the redwoods, people have a love of natural places that is deeply ingrained into their very being. The love of these places and the deep desire to protect them lifts the looming environmental catastrophe from an abstract fear to something that we can act upon. Simple actions can sustain and protect these places: being mindful of our use of scarce resources; participating in groups working to save land; and supporting government actions to protect these places. When multiplied around the globe, these individual and collective actions to save natural places will protect the green infrastructure that sustains life on earth. But perhaps as importantly, these actions will create a network of places that continue to inspire people and give them a personal reason to act mindfully. This brings me back to the redwoods. I know that however I live my life right now, an 1,800-year-old tree nestled in the heart of Humboldt Redwoods State Park is quietly recording the impact my decisions have on the atmosphere that sustains us both.

❧

Ruskin K. Hartley is the Executive Director and Secretary of Save the Redwoods League, a nonprofit organization that protects and restores redwood forests and connects people with their peace and beauty so these wonders of the natural world flourish. Hartley, who has served in his current position since December 2006, is the sixth leader in the organization's ninety-one-year history. In his previous role as Director of Conservation and Education, he developed the plan for the League's conservation efforts. Visit www.savetheredwoods.org.

CHAPTER THREE

Taking Single Steps

Dusting Off the Energy Solution
in the Basement

DANA GOLDMAN

It's cheap to maintain, highly energy efficient, and is probably in your basement, collecting dust. Until a few years ago, that's where my clunky Schwinn knock-off had been since high school, mounted up on a wall since I'd discovered, at age fourteen, that bikes are only for the very young, people who wear Spandex, and all of Holland.

Of course, that was before gas prices went through the roof and then doubled; before my first car became the high-maintenance automobile equivalent of Zsa Zsa Gabor; before theories about an end to oil started circulating more than the wine and cheese at a cocktail party.

This before-time now feels further away than the manmade end-times that now seem inevitable. There's the important question of which scenario will win (Will it be the scarcity of gasoline? Global warming? Food insecurity?), but an answer is a moot, if interesting, point: any one of them has better odds than the continuation of our same-old existence.

And so, broke from my lemon of a car and scared of the small talk centering on oil, I returned, humbly, back to the basement to wipe down that two-wheeler I so recently scorned. Neither very young, nor Spandex-clad, nor European, I began to ride.

"An experiment," I told my dad, who loves bicycles as long as the people who ride them in traffic and rain and in the dark are not his daughter. "If it doesn't work, I'll buy another car."

But, a year and a half into this experiment, there's no incentive for me to stop. Living almost all by bike, I've come to take on the U.S. Postal Service's motto, if not its vehicles or rates, as my own: neither snow nor rain nor heat nor gloom of night stays this courier from the swift completion of my appointed rounds. There's nothing like a false sense of heroism to get me up hard hills. And I can fudge the swift part.

Much of bicycling is hard. Besides hills, there's weather and potholes and cars to contend with. But I wouldn't have it any other way. Bicycling doesn't let me off the hook from the galaxy of impending environmental disasters that seem headed straight to earth, but the clicks of simple gears and the deep breaths I take coasting downhill reassure me there is a plan B, and that I am living it. It doesn't matter as much if gas becomes unaffordable or disappears like the dinosaur; riding my bike means I can ignore short-term stuff like car payments and be at peace for the long-term about how I'm going to get around. Beat that, post office: neither global warming nor an end to oil will be able to knock me from my twelve-miles-an-hour pace.

More difficult than the actual biking has been the disbelief from car-commuters, also known as my friends and relatives. Before gas prices went up the question I got most was Why? (Lately I think car-commuters have been answering that one themselves.) Now they're on to a new one: how do you do that?

In fact, the how has been the easiest part: I move one pedal and then the other, and keep going. I pack rain gear, just in case, and have slowly learned to take advantage of big bike bags and what's within my biking radius: the close grocery store and eye doctor and bank, the flatter routes and quieter streets.

The more uphill battle has been trying to convince those who ask How? that it's not just me who's capable of pushing down one pedal, then other, and eventually ending up somewhere. Spinning classes and Lance Armstrong have put recreational bicycling on the map, but bicycling with a full load of groceries in work attire still challenges our ideas of business—or cycling—as usual.

What's left mostly unsaid when people ask How? is a statement of belief: "I could never do that." It's the same kind of self-doubting and self-defeating feeling I had when I first began to model myself on the P.O. We believe we are not strong in the bike-to-work sort of way; we know we do not have or like or look good in skin-fitting clothing, we are overwhelmed by the logistics of such a venture.

But our beliefs about ourselves sometimes get in the way of who we are and who we are becoming. These days, with my bike routinely dusted and its tires properly pumped, I know that we do not have to be strong to begin. We do not have to feel invincible to make it from location A to location B. We do not have to be young, or European, or look good in Spandex. We don't have to know exactly what we'll do when it first begins to rain, and we don't have to keep on biking when we're tired and on an uphill.

We merely must be willing to begin: to begin to see another way and to be willing to see ourselves in another way; to keep moving through whatever comes, be it drought or self-doubt or uphills or downhills. And so with my bike carry-ons, I carry on, delivering the messages of pleasure and freedom that come bungeed to this forgotten choice: adapting, shifting, coasting, and preparing for a sea change as, pedal by pedal, I find myself moving once again.

❧

Dana Goldman has worked as a writer, public radio journalist, teacher, and wilderness guide. Her work has appeared on National Public Radio, in the *Atlanta Journal-Constitution,* and in the anthology *Quarter Passed.* She lives in her hometown, Atlanta, and is now studying to be a licensed professional counselor. In between classes, she facilitates outdoor explorations with her husband and his company, Sure Foot Adventures (www.surefootadventures.com).

Every Day We Choose

FRANCES MOORE LAPPÉ

Living fully during a time of historic crisis for our planet is possible, I believe, only if we are able to grasp how our individual choices address its very roots. The planetary social and environmental catastrophe we face can feel overwhelming. But I've learned that even when a task seems *huge*—cleaning out the attic or writing a book!—I do find the energy to tackle it *if* I can see first steps. *If* I can see how a small action—getting together a few boxes or creating a one-page outline—connects to my ultimate goal, an attic where I can actually find things, or a book that might help me find answers. I feel overwhelmed until I have an idea of how to get started and a picture of how it will add up to something.

The same inner experience holds true as we turn to much bigger challenges. To connect our passions with the world's needs in ways we sense really do "add up," we must probe deeply enough to see the underlying patterns trapping us in this horrific mess. I believe we can then stop repeating the "same-olds" and expecting something better to happen. Grasping causal patterns, we can feel excited—not loath—to change.

How can I interrupt a negative cycle that creates suffering or reinforce a positive one that contributes to new, life-serving rules and norms? That's the question. And to answer, it's helpful, to me, to distinguish between "issues" and "entry points."

"Issues" overwhelm. They hit us as distinct problems, piles and piles of them. We hear of child slavery, violence against women, hunger, of HIV/AIDS, deepening inequality, pollution and global heating, depression, failing schools ... I feel buried, smothered under a mountain of problems. I want to cry uncle.

HOPE BENEATH OUR FEET

"Entry points" are very different. Entry points we can detect because we're weaving a theory of causation. So we can pinpoint places to start to shift the killer cycle itself. On the surface they might appear as distinct problems, but they are ways "in": they are sharp points that break into and deflect the downward spiral of powerlessness; they are deliberate actions that strengthen the flow of positive causation, putting in motion an upward spiral of empowerment.

To make these distinctions clearer, let me give you one example: "power shopping." Which head of lettuce you pick up today or where you buy your next T-shirt may not seem like a world-changing decision. But it is.

Spirals of powerlessness are generated not only by laws on the books but by norms that our daily acts create. If we buy pesticide-sprayed food, we're saying to the food industry, yes, yes, give me more of that. If we buy organic instead, we are stimulating its production. (Why do you think McDonald's serves organic milk in Sweden but not here?) True, these marketplace "votes" are grossly lopsided—for the more money one has, the more votes one gets—but our purchases make ripples nonetheless.

I say this not to make us feel guilty but to help us realize our power.

Sixty-three million Americans now say they base their purchasing decisions on how they affect the world, and four out of five say they're likely to switch brands to help support a cause when price and quality are equal. Even ten years ago this was hardly the case.

Worldwide, sales in the Fair Trade movement jumped by over 50 percent in just one year, 2004. It now functions in fifty countries because millions of consumers are seeking out its label, guaranteeing that producers receive a decent return. Just to take one example of its impact: in 2006, Rwandan coffee cooperatives (whose members include widows and orphans of the 1994 genocide) received a Fair Trade price for their coffee that was three times higher than that offered by local merchants.

This sea change in awakening to the power of our purchasing choices comes to us also thanks to some energetic, determined people. One is Lina Musayev. Now twenty-five, Lina was a student at George Washington

University when her life changed forever during a 2002 Oxfam America leadership training intensive.

"Farmers from Guatemala came to talk to us," Lina told me. "We got the real story of Fair Trade from the roots. I didn't know anything about the coffee crisis. I didn't know it affected twenty-five million people. So when I heard about Fair Trade, I thought, 'This is incredible. It's working. It's making a difference.'

"The next day, literally, my friend Stephanie [Faith Green], who'd come with me from Georgetown University, and I founded United Students for Fair Trade.

"She and I are really close. We made a great team.

"Once school started, I decided to start from the bottom with a petition saying students wanted more Fair Trade coffee, and we got two thousand students to sign. That's out of ten thousand. It worked. We sent a letter to Starbucks. We pushed for Fair Trade coffee at every university event, like teachers' meetings."

I asked Lina what approach she'd found most effective in reaching students.

"The main thing is getting farmers themselves to come to the campus. Hearing the farmers, I see the students say, 'Oh, my gosh—I didn't know this.' Almost like I was!"

After three years, George Washington passed a resolution that called on all on-campus vending outlets to serve 100 percent Fair Trade coffee. In only five years, the student Fair Trade movement Lina and her friend Stephanie launched has spread to three hundred campuses, and roughly fifty campuscs now serve only Fair Trade coffee.

Lina and Stephanie would probably find it hard ever again to view economics as simply about things exchanged in anonymous transactions. They are helping shape a new norm, an economy that's about people, people relating with each other—fairly.

Joyful living, I'm convinced, happens when we hit that spot where a potent entry point that touches root causes fires our own deep passions. I know that when I first discovered that spot—my mid-twenties' "aha" that our daily eating habits make huge ecological and fairness ripples—it set off a personal revolution, and I've been forever grateful.

To find that spot, a critical first step may be to recognize that the negative spiral can start deep inside us. If a feeling of "lack" lurks at the center of our pain, pain that we then project out and create in the world, we can start within ourselves to reverse it: we can acknowledge sufficiency. Right now, we can focus on the strengths of ourselves and our loved ones and the possibilities in front of our noses to enhance our capacities and meet our needs for fairness, cooperation, efficacy, and meaning.

Awareness of these capacities can propel the spiral of empowerment busting us out of any downward spin.

So think of something you are doing right now. Maybe you are engaged in your children's schools to make them more empowering for students, or you're sending off an email to the newspaper shaping your community's views. Maybe you've chosen to lighten your weight on the planet by eating less meat, converting your home to solar energy, or joining in "community-supported agriculture" by buying a share in a nearby farm's produce. Maybe you are finally speaking about discrimination you see in your workplace or going door-to-door on behalf of a candidate who is actually listening to citizens' concerns.

Think of what you are doing, and then think about what you have always wanted to do. And ask yourself: am I expanding and spreading power? Am I easing fear of change and fear of the other? Am I learning and teaching the arts of democracy? Am I creating a sustainable movement? Am I replacing the limiting frame with an empowering one? Ask yourself these questions, and believe that change is possible.

I admit it: in the 1970s I never could have imagined the world as I experience it today. I assumed things would get better (if people listened to me, of course!); or they would get worse. But, it hasn't turned out that way. Things are moving fast in two directions at once: they are getting very much worse *and* they are getting very much better. The real challenge is staying sane in this both/and world: it is holding both realities.

It is not possible to know what's possible. This is how I now understand humility. Believing we can accurately predict outcomes, as cynics claim to, has become for me the utmost in hubris. And because this is true, we are free. We are free to act assuming that our action—no matter how "small" it appears to us—could be the tipping point setting off tectonic shifts of consciousness and creativity.

We cannot predict outcomes, but some things are coming clear, and that clarity is beginning to rattle us: the shock of melting ice caps and dying penguins, of leveled rainforests and species wiped out daily before we've even met them, of children armed in genocidal war, and children dying of hunger while more than a third of the world's grain goes to livestock . . . all of this is sinking in, and more and more of us know the time is now—that we act powerfully now or we see our fate sealed. We risk becoming our species' most shameful ancestors, passing on to those we love and those they will love a diminished world that we ourselves find heartbreaking.

Such shock may then open us to the surge of energy lying dormant—that pure, protective rage we can transform into exuberant defense of our beautiful earth under siege.

Yes, there is much we do not and perhaps cannot know about our chances of success. But there is much we *can* know:

Humanity is coming to understand nature's fundamental laws and the fatal consequences of ignoring them. Rather than triggering panic, though, coming to accept nature's boundaries may bring huge relief. If children need boundaries to feel safe, maybe we'll find we all do. Nature offers us real, non-arbitrary guidelines, and as we align ourselves with her—because we ourselves are part of nature—we may also move into greater alignment with one another. Could this shift, truly trusting nature's laws, ultimately release the grip of self-created scarcity, allowing us to experience real abundance for the first time?

This takes courage, and courage is contagious. Many are also coming to know that just as we need not fight the natural world, we need not fight our own nature. We can trust our deep, in-born needs to "connect and affect." We can trust our ability to walk with fear. We can even trust our capacity to let go of long-held ways of seeing in order to structure our societies to bring out the best in us while protecting us from the worst.

Ultimately, if we accept ecology's insights that we exist in densely woven networks, then we must also accept that every choice we make sends out ripples, even if we're not consciously choosing. The choice we have is not whether, but only how, we change the world.

⤟

Frances Moore Lappé is a democracy advocate and world food and hunger expert who has authored or co-authored sixteen books. She is the co-founder of three organizations, including Food First: The Institute for Food and Development Policy and, more recently, the Small Planet Institute, which she leads with her daughter Anna Lappé. In 1987 she received the Right Livelihood Award. Her first book, *Diet for a Small Planet,* has sold three million copies and is considered the blueprint for eating with a small carbon footprint since long before the term was coined. For more information, visit www.smallplanet.org and www. foodfirst.org.

With the Turn of a Key,
I Can Make a Difference

One fine afternoon last spring I was sitting in my old Volvo waiting for my son to get out of school. I was a bit early, so I'd parked in my favorite spot, with my windows open to the woods and a view of the school entrance, and settled in with my journal to enjoy a bit of peace and quiet. Just then, a spotless white Escalade pulled up next to me like a cruise ship docking, and I noticed my mind formulating an instant story about my neighbor. As I waited impatiently for her to cut her engine after parking, my journal entry became a rant about her and all the other people who buy, and then idle, their SUV's. *How can she idle so thoughtlessly, as if her actions made no environmental impact, as if her vehicular living-room with its TV and air-conditioning completely insulated her from the impending global warming crisis all around her?* For fifteen minutes we sat side by side and worlds apart, she on her cell phone, and me rehearsing but not delivering a strident and holier-than-thou speech about why she should turn her engine off (and also completely change her entire lifestyle). After that, I started to scan the school parking lot every day at pick-up time, and noticed with dismay that ours seemed to be a culture of SUV idlers. I made up a story of resentful isolation, as if I were the only person in the whole school community that cared about the environment. I sat fuming in my little car, consumed with fearful images of the future of our planet and discouraged by my own apparent inability to make any significant change to improve the situation.

In retrospect, I'm glad I didn't approach any of my fellow parents last spring because I wasn't ready to communicate clearly and peacefully. No doubt I'd have transferred my simmering cloud of anxiety about global warming into a blistering lightening-bolt accusation that would have put anyone on the defensive. Even if I tried to mask my scorn and self-righteousness with polite affect, knocking on the window of a Hummer to tell the driver to turn off her engine would certainly have backfired if I believed she was a perfect example of wanton overkill consumerism and deliberate, head-in-the-sand ignorance. What did I know of her life, her family culture, or the conflicted justifications for choices she might have made? The only person I can hope to understand is myself, and the discomfort I felt around the dilemma of how to take effective action was a sure sign that I needed to find a different way to frame this issue.

Since then, I've arrived at this analogy: if it is so easy for me to believe that others should turn off their engines, why can't I turn off my internal "judgment-engine"? Who is to say which is ultimately worse for the environment, the Escalade's exhaust or my toxic rant about the behavior of another human? What if these people I'm judging simply believe what they've been taught and they don't yet understand the consequences of their actions? It is likely that they feel they are being responsible, caring parents to come and pick up their offspring, and keep the inside warm or cool for their comfort, and they don't realize the extent to which they are poisoning the air these children breathe as they walk out to meet them. I didn't know myself until recently that it isn't necessary to warm up a car's engine by running it for a long time in winter, and I didn't realize how much gas is used when we idle our cars, until I did some research and found out, for example, that *an idling vehicle emits twenty times more pollution than one traveling at thirty-two miles per hour. One hour of idling burns up to a gallon of fuel and produces approximately twenty pounds of carbon dioxide. Children breathe 50 percent more air per pound of body weight than adults, and are thus much more vulnerable to airborne toxins. More than ten seconds of idling uses up more fuel than restarting the engine. Since 1972 it has been against the law in Massachusetts to idle unnecessarily for more than five minutes. This past year, a new bill entitled "an act to improve school campus air quality" has been passed which prohibits buses and commercial and personal vehicles from idling on school property.*

How would I approach other drivers if I assumed they really wanted a safe, clean world for themselves and their children but needed a gentle, compassionate wake-up call or simply some good information? Even more radical than that, how would I approach them if I didn't need them to change at all? How peaceful would I be if I could accept them and myself exactly the way we are, and still take action? I realized I had energy around this issue and I felt compelled to share what I'd been learning, for my own sake. I simply couldn't sit back and be passively frustrated and judgmental any longer. For me, the best course of action was to launch an idling-awareness campaign and to learn how to give out the information with an open heart.

Now I am helping to put up "Idle-Free Zone" signs around the town's public schools to remind parents and bus drivers not to idle while waiting for kids. I've made up an idling fact sheet to send home to parents, and I'm hoping to involve the science teachers so that they can incorporate some emissions information into their curriculum on Earth Day. When I presented this project to my son's principal, he not only approved it gratefully but also volunteered to walk around after school handing out the flier to drivers who appeared to need a more direct, personal approach. For my own neighborhood, I've adapted a more generalized fact sheet to give to the taxi drivers and take-out delivery people who idle outside of my building. I've proposed to the high school's environmental club that they target the local drive-through restaurants and hand out fliers to people who are idling there, urging them to consider parking instead of crawling along in the drive-through line. Every day I get to practice my approach, and it feels most authentic and clear to me when I present it in the spirit of sharing a true gift, from one human to another. It is information about a simple shift that can be made, something one can do that will help us all out. This feels immeasurably more constructive than what I was doing last year: sitting alone in my car, fuming to myself and idling my own "judgment-engine." I'm grateful that I found a way to turn that key.

Neige Christenson is an improvisational dancer, teacher, therapist, writer, and mother. She is a graduate of the School for the Work of Byron Katie, which has supported openness and creativity not only in her art but also in her relationship to community and to the environment. Her writing has appeared in *Contact Quarterly, The Sun, Fire in the Womb: Mothers and Creativity* (Xlibris press), and *The Mom Egg,* a literary periodical of Mamapalooza. She lives with her family in Boston, Massachusetts.

Shut Up and Vote

ERIC RUBURY

After thousands of years of mankind's self-inflicted wounds such as war, poverty, slavery, and genocide, who would have thought we could come up with a new one like global warming? What to do? How to think about it? Well, we can always whine and shift the blame: SUV soccer moms. Capitalism. Television (except the good shows that I happen to watch, including "Monster Truck Makeover"). The Other Political Party. Maybe some collective guilt, such as gravely intoning that "we're all responsible." And although some of these easy answers are crowd pleasers (some have actually made me look smart at dinner parties), it doesn't accomplish much—sort of the sonic equivalent of graffiti. And at this point, isn't criticizing global-warming-deniers a bit like shooting fish in a barrel?

Besides, I try to restrict my high-moral-ground whining to a minimum, and to only what is necessary to impress my peers, friends, family, and neighbors. I try to restrict it because . . . I am a hypocrite. I was once part of this global warming problem in a big way. It turns out that I've had to work for a living, and my chosen vocation at one time in my life was that of a geologist. An oil company geologist. A Big Oil company geologist. It gets worse, so hang in there: a Big Oil company geologist who prospected for oil in some of the most environmentally sensitive places on this planet. Real sensitive places. Animal Planet–National Geographic–Discovery Channel kinds of places. So many wells that I sometimes lie awake at night and can almost hear the whiffle ball-like sound coming from the holes out there.

I now realize I was part of The Problem. My first awareness of complicity was on a well site in Guatemala. Guerrillas had attacked the rig, executed a handful of people, blew up the generators and landing strip, and no one was going to secure our jungle site for at least a week. I found myself hiding at the edge of an open waste-oil pit and thinking two things: 1) I need to ask for a raise, and 2) How am I going to explain my role in this mess to our children years from now? Staring at a seat in Hell impels one to take stock of life and think about what one can do in one's remaining years to perhaps be eligible for some sort of parole. Fortunately, dear reader, you likely don't need to worry about these things—unless you drive a car (and yes, a hybrid is a car, so you still get a full demerit); heat your home with gas, oil, electricity, coal, or wood; or buy organic salad greens in plastic containers (recycling only saves you half a demerit). And being guilty of any of these makes you an enabler. For instance, by buying tomatoes in plastic wrap, in your own small who-me? way, you have enabled the plastic manufacturer, the food distributor, and the supermarket by signaling to them that it was all kinds of acceptable to produce such products in such a fashion. And that gas for your car—it was delivered at such low prices that it helped fuel both you personally and the unprecedented growth in the world's economy for the last century. Yes, you're welcome.

And although I ended up finding very little oil personally (small comfort to either the environmental cause or my former oil company bosses), I still search for ways to be a better citizen in the last half of my life than I was in the first. What I've found is this: the effort that takes the most amount of actual work (recycling, composting—I try to do it all now) might make me feel a lot better about myself, but it's the small, effortless piece of work that has the true potential to change the outcomes of global warming and looming environmental catastrophe. It's called voting.

Voting has a bad rep on the Personal Responsibility List exactly because it is so easy to do, and well, hey—it's only one vote. But I ask you to consider this: what has the greater chance of creating real change in this world, your compost pile or the installation of progressive, rational, and influential political leaders to run this scene? This is a great nation— we have won world wars, invented the Internet, jazz, baseball, and peanut butter cups, and put guys on the moon. This nation and its people can do big things, do them successfully, and help the rest of the world achieve important goals. We have the money, we have the technology, and we have always been a nation that accepts personal sacrifice for the greater good: we're set to roll! But we need our government to get with the program. Not an easy task, and not something that I have the perfect answer to. But someone does, and ideally they are ready to risk their political capital on successfully addressing global warming. If not, they don't get my vote, and they shouldn't get yours, either.

Readers of such a wonderful book as this, and especially those who find this particular essay intelligent and thought provoking, probably are good citizen-voters already. While such an assumption comes a little too close to blaming the low voter turnout on people-other-than-you-and-me, let's accept that premise, at least for the remainder of this paragraph. In the last presidential election, 121 million Americans voted. Let's assume that only 10 percent are really concerned about the environment and global warming (polls suggest that there are many more than that). And let's say that only half of *those* are really, really, really concerned, enough to go out and convince one—only one—person they know who either regularly votes for the Other Party or doesn't vote at all to vote for the candidate with the most progressive and aggressive plan to save this planet. We're talking about only the most concerned, wild-eyed progressives here, just convincing one—one!—voter each. If that happened, everything would be different. Even the supposed difference between Red and Blue states amounts to a couple of percentage points either way. It's that *close,* people. Most of our dinner party chats involve talking with people who think like us. Find someone who doesn't. Talk to him or her. Go back to the cheese platter and wait; someone will show up. But if you think that the voter-turnout problem lies somewhere else, to paraphrase the great comics character Pogo, we have met the enemy and he is our friend and neighbor. And perhaps one's self as well.

Researchers have found evidence of recycling and composting by our ancestors during Paleolithic times. Look what good it did *them.* And yes, recycling and composting, along with driving fuel-efficient cars, are great and worthy personal actions. True voting in a democratic society has only been around a fraction of that time (again, one of America's inventions), and it has achieved great things.

There's an old adage that states that the smaller the deed, the greater its effect. At least I *think* it's an old adage. It certainly *sounds* like one. At any rate, I'm not an expert at proselytizing, but I have a few sins to compensate for, so I'm making good on my promise to change the heart of one voter—in fact, I even know which one. There have been way too many close votes in my lifetime, and I'll be darned if I'll add personal voter apathy to my list of failings. The same, alas, should go for you, too.

If there's anything you take away from this little rant, please make it that this is a great nation, with great people, a great history, and great potential for the future no matter what it brings; and that good citizenship mandates not giving up hope for the institutions that have done so much good in the past. That, and that all sinners—like me—have futures.

❧

Eric Rubury lives in Redding, Connecticut, with his spouse, Leslie Cohen-Rubury. He is currently Chief Financial Officer for OceanConnect, which, among other business lines, includes the promotion of renewable biofuels and personal carbon credits. They have three children and a future daughter-in-law. He's lead guitarist in the local rock band Stuntfish. He has worked and traveled extensively in Central America, South America, and West Africa.

One Piece of Paper

KRISTINE ALACH

We stood and stared at each other blankly. The rotted smell of decaying lunches rose up from our recycling bins. Straggling students wandered past on their way home for the weekend. Teachers trudged by, and some woke from their tired trance to acknowledge us before walking outside. Even though the work and school day had ended for some, our work had just begun.

In fact, the work that led to a recycling project at Oliver Ames High School in Easton, Massachusetts, had started about two years earlier. My best friend Lydia and I always had a thing for the environment. She liked learning about it; I liked rolling in it. We both would rant about the importance of conserving energy and protecting the earth. Any new information we gleaned from school, magazines, newspapers, the Internet, billboards, or television shows was quickly catalogued and became artillery for our environmental debates with any willing party.

We talked about the need for change at school. Mostly the school staff just talked, so we began our own private recycling. We started with our own bottles and eventually we accepted donations from friends. Lydia would bring the bottles home to recycle. However, it got to a point when we were sitting on the floor of the hallway trying to find some secret pocket that could possibly hold just one more bottle, since our bags were already stuffed full of plastic. It was then that we realized that if no one else was going to make the change at school, we had to.

We soon discovered that many people wanted the school to recycle, but none wanted to invest the effort to make it happen. There is something to be said about the philosophers who go about prophesying and weighing all the pros and cons of every given situation. However, sometimes you have to attack a situation head first, holding onto your passion and goals. If no one had ever taken the initiative to improve our world, we would still be picking bugs from our hair around a campfire dressed in loincloths. Part of being human is being compulsively curious about new ways to improve the improvements of yesterday. In this spirit we set out to lessen our school's eco-impact.

We dodged political bullets, ducked under red tape, wove through obstacles, danced through countless phone calls, and tap-danced along trouble. Through the year we brainstormed ways to organize the people power (work force) we would need to manage our collections. No amount of preparing could have made us ready for the challenges we would face.

Previously our school had thrown away paper and plastic, along with our lunches and broken pens. Changing this took an effort from everyone at the school, and involved redirecting the janitors, creating collection teams, informing teachers, training students, and generally creating new habits to replace the old. (Many teachers mentioned that even when the trash bin was next to the recycling bin, they found it difficult to remember to separate the two.) With a little effort from everyone, our recycling results have exceeded our greatest expectations, chipping away at our wastes.

A small group of five students had signed up to dedicate their afternoons to collecting recyclables at the high school, and we were soon elbows-deep into our jobs: sorting paper and plastic, emptying bins, and returning rolling totes. However, this was no easy task. With more than a thousand contributors to the recycling effort and over a hundred bins to collect, there was a big margin for mishap.

Since then, we have encountered some predicted problems and some that were unforeseen. We have to deal with cleaning out trash that is mixed with the paper and plastic, including countless lunches and construction materials. One day we found a whole chocolate cake upside down in the bottom of a seventy-gallon rolling bin. Needless to say, with no way to reach the bottom of the barrel without climbing in, it was an exercise in problem solving.

Days like that one make me wonder if all of this work is worth it. How much does recycling mean to me? Is it worth giving up my free time to clean and empty bins and make phone calls? Then I just think about watching our recycling dumpster grow full of paper week after week. I think about how teachers and students alike comment on how they are ashamed that they did not start recycling earlier. I think about how fellow students and parents watch us each day sorting and dumping paper into the dumpster, and how they have become more aware of the problem through our work. Finally I think about the blades of grass sprouting around the dumpster and the feeling of sunshine on my face, and I think that maybe I can help my grandchildren feel the sun and enjoy our earth.

When I doubt myself, I wonder how much of a difference we really make. Then I think of one piece of notepaper that is put in a recycling bin by one student, and all of the paper from that class piled on top. Then how that one bin is combined with all the contents of the bins on that floor into a rolling tote. Then how all the rolling totes are dumped into a large dumpster and how soon the dumpster is full, and how the dumpster fills every two weeks. All of that paper to be recycled, and all of it starting with one student and one piece of paper.

≈

Kristine Alach currently attends the University of Massachusetts, Amherst, as an Animal Science major. A passionate equestrian, she loves to work with horses and study their behavior. She recognizes the need for change in many facets of the world.

A Five-Hundred-Year Plan

JANE HAYES

A while back I met a military man who had a fifty-year plan for his life, career, and family. Our conversation became lively when I admitted that I've been thinking about a five-hundred-year plan and it has lots to do with beans.

My plan is inspired by the Iroquois edict of caring for the welfare and well being of the seventh generation to come. That means I care for my grandchildren's grandchildren's grandchildren.

It helps me to hold great great grandchildren in mind as I face the challenge of changing my untenable cultural habits. Until I had them in mind, I hadn't been able to imagine effectively acting on daunting issues like war, global food security, and climate change. Now I can let go of immobilizing feelings that we are never going to be able to get past these issues peacefully and intelligently.

My gut tells me that it is time to lighten things up a little. No matter how dire the situation, we cannot live on fight-or-flight our entire lives. Our intelligences are reduced by 20 percent when we are in a stressed state. Exhaustion doesn't help our chances either. We need spiritual fuel, loving connection and engaged discussion about what to do and how to organize our actions, big and small. We need to start with where we are, personally and in relation to other living beings.

My "avant gardening" friends are most diligent in thinking about seed saving for the future. We know we need food, seeds, and cooperation to keep human presence on this planet. Earth knowledge and seed viability are being lost at staggering rates. Going with my personal love of food and gardening, and my drive to invite people to take responsibility for the future, I've chosen to invest my energies in tackling food security issues with a long view. I am grateful that others are investing in growing and saving as much rare seed as possible and tending organic farms. In turn, I'm reaching as many people as I can with stories that share some early blueprints—some small piece of a five-hundred-year plan.

The Bean Keeper story (for children ages seven to ten) is about six children in a small Canadian town who learn that the crops have failed due to drought. The adults are worried and the town's harvest festival gets cancelled. The kids decide they'd like to help by growing the only thing they know how—beans. Luckily, they had learned how in kindergarten, like millions of children in North America. The kids collect beans and eventually meet old farmer Joe, the only Bean Keeper left in their county. He gives them dozens of varieties of beans and teaches them how to grow them and save the seed. The kids are excited to start, but face a long winter ahead. They put on a school play about being Bean Keepers to pass the time. By spring the rest of the school wants to join in and a hundred kids are organized to grow a hundred kinds of beans. The kids inspire the adults with their efforts and the harvest festival is reinstated.

I told the story orally for a while and was surprised when I began to meet people who had heard the story already. Something is catching on. A few organizations have decided to help. Seeds of Diversity offers free beans to groups that want to be Bean Keepers. Evergreen hosts the story online and supports Bean Keeper projects with resources, funding, and expert assistance. Evergreen reaches sixteen thousand schools in Canada and Seeds of Diversity estimates there are a few thousand beans that will grow up here. It isn't hard to imagine stewarding thousands of beans in time with this kind of interest.

As schools and groups from across Canada sign up to be Bean Keepers, a set of grade three lesson plans linked to Ontario curriculum has developed. We've added in some of the lessons that old farmer Joe offers. We've written and recorded the Beans song with Jerome Godboo's blues band. We are testing school and festival level performances of the story with the help of school eco-literacy programs. As the story matures in community contexts, more players are coming to the table. Farmers are sharing their bean seeds and knowledge with schools. Cooks and chefs from a variety of backgrounds are invited to make Bean Keeper dishes. School children are invited to write to each other. At the Gardiner Ceramics Museum, students can make clay bean pots and storage vessels while learning about paleontology's late discovery that five-hundred-year-old beans can sprout if seeds are saved in the right conditions.

I'm five years into my five-hundred-year plan and there are another twenty lifetimes past mine to figure out. I'm optimistic. I have faith in humans to creatively move past this moment, regardless of how hard it is. I think we will all benefit from diving into the work of knowing the earth arts, pushing ourselves to be athletic in our stamina as we save knowledge and seed. We should take breaks and celebrate too. Without these, we become soul-tired, sad, and lackluster. Breaks and celebrations, music making, and sharing food are an integral part of the plan.

⤳

Jane Hayes is a gardener, artist, and social entrepreneur. During her career, she established the award-winning children's garden program for the City of Toronto, helped launch the Toronto Community Garden Network, and contributed to Evergreen's national school ground greening program. In 2007, she founded Garden Jane, offering garden education and recreation for people of all ages. One of her favorite projects is Cirque Dirt, an earth circus focused on holistic community building. Jane holds a bachelor's degree in Anthropology, a Masters of Environmental Studies, and is certified to teach permaculture. Visit her on the Web at www.garden jane.com.

CHAPTER FOUR

Little Steps to Big Leaps

Fight It Head On

BILL McKIBBEN

I'll tell you how I live my life right now. Right now I'm on the train from Berlin to Munich. Last night I gave a talk in Berlin, after one in the morning in Copenhagen. The day before that half a dozen meetings and speeches in London. In the next week I'll be in Sweden and Switzerland and Turkey and Portugal, then on to India, the Maldives, Egypt, Jordan, Lebanon. Etcetera. I live my life right now in constant motion, organizing 350.org, the first big grassroots global campaign to demand action on climate change.

As I write this, we're four months out from our big day of global action in October 2009, designed to influence the Copenhagen conference in December of 2009. We don't know yet if we'll actually have any impact, but we do know the day itself will be a huge success—there will be mountain climbers high on every range, 350 scuba divers on the Great Barrier Reef, big demonstrations in cities across the globe. In Delhi thousands of people will form a giant "3," and in London a "5," and in Copenhagen a "0." Why 350? Because in the last two years, since the melt of Arctic ice in the summer of 2007, our finest scientists have made it clear that's the most important number in the world—it's the most carbon we can have in the atmosphere and maintain a planet "similar to that on which civilization developed and to which life on earth is adapted." And since we're already past it, at 389 and rising, that's bad news.

So: I've got no very complicated or interesting answer to the excellent question this book raises. I've been thinking about it for twenty years, since I wrote *The End of Nature,* the first book for a general audience about climate change. I've spent time doing important things—trying to seriously pay attention to the natural world, which will never be as intact as it is now. Trying to work in my community in the mountains of the Northeast to build the kind of institutions and economies that could both slow down climate change and help us adapt to that which we can't prevent. And most importantly, I've tried to raise a resilient and grounded daughter (the task where I've met with the most success, though I probably had nothing to do with it at all).

But—more and more—I've spent my time organizing. First a march across Vermont, which turned into what the newspapers called the largest climate change demonstration in U.S. history. Then, side by side with six college kids, a campaign that coordinated 1,400 simultaneous rallies across the U.S. Then, with Wendell Berry, I issued the call for the first mass civil disobedience on climate change in American history, a day coordinated by groups like Greenpeace and the Rainforest Action Network that actually persuaded Congress to shut down the coal-fired boilers that used to provide electricity for the capital. Now we're working on a global scale on 350.org, with offices across the world, mostly run by incredibly dedicated and talented young people. In Delhi, in Beirut, in Johannesburg, in Quito, in Budapest.

I fear that I've left behind much of what used to define me. I was an essayist, a writer, a thinker. Much of that subtlety has been stripped away, perhaps for good. Cornered by this greatest of troubles, I couldn't think of anything to do but try and fight it head on. So I did. And I'm glad I did—the rewards have been worth the cost, though the cost has been real. For the first time in twenty years I feel as if I'm doing exactly what I should be. I may wake up in the morning worried about the world, but also knowing that I can do something about it.

Probably not enough to win. The science is very dark—when the Arctic melted in 2007 it became clear that we'd crossed a threshold much much sooner than we'd imagined. The sheer momentum of the warming is such that even if we do everything right from here on in, there's no guarantee that we haven't set forces in motion that will convert this sweet earth into a different planet altogether. Still, our work changes the odds a little. A 20 percent chance isn't all that good, but it beats 15 percent. Sometimes the dice roll the way you hope.

And there's something more about this 350.org project. We're taking this new fact about the world—arguably the most important fact about the world—and trying to drive it deep into the imagination of the earth. In the hopes that it will help reset the psychological bar for the negotiations—but even if it doesn't, it will at least let everyone know what the reality is. That task, it seems to me, is arguably literary—I'm writing with a crayon, but I'm writing, trying to push the actual state of the world into the various lush and corroding fantasies that now control our public life, and often our private too.

It's not very complex—I've been forced to realize I'm not a very complex thinker. I'm down to three digits now.

❧

In 1989 Bill McKibben wrote the first book for a general audience about global warming, *The End of Nature.* He has since authored a dozen books and led the largest national and global campaigns about climate change. He is a scholar in residence at Middlebury College. His Web site is www.billmckibben.com.

Become an Urban Homesteader

KELLY COYNE and ERIK KNUTZEN

Prompted over the past few years by oil wars, global warming, ecological collapse, natural disasters, and our psychotic federal government, we've made a few changes in the way we live.

Now the day begins when Erik gets up to let the chickens out of their henhouse. It's a structure so thoroughly secured against marauding raccoons that we've named it "Chicken Guantanamo." The hens have been patiently waiting for that door to swing open since first light. Next, while the coffee brews, Erik throws some flour and a cup of sourdough starter into the mixer. He bakes a loaf of artisanal sourdough bread for us every other day, and we rarely meet with any bread that tastes better.

I get up a little later than Erik and stagger into the garden first thing. I say hi to the hens, add some kitchen scraps to the compost pile, and turn on the drip irrigation systems that water our vegetable beds. As of this writing our garden is bearing tomatoes, cucumbers, fava beans, Swiss chard, figs, ground cherries, leeks, eggplants, assorted herbs, and a selection of cultivated weeds. I'm looking forward to the corn, avocado, and pomegranate harvests, all of which are a few months away.

For breakfast I enjoy homemade yogurt with raw honey or maybe a thick slice of the aforementioned sourdough, toasted and smeared with tangy homemade apricot butter. After breakfast I take three sheets of tomatoes down to the solar dehydrator so we'll have sun-dried tomatoes in the winter. Then I hang a load of laundry out on the line.

Where do Erik and I live? In the heart of urban Los Angeles, in a decaying bungalow on a small plot of land. We are urban homesteaders.

What Is an Urban Homesteader?

An urban homesteader is someone who enjoys living in the city, but doesn't see why that should stop her from engaging directly with nature, growing her own food, and striving for self-sufficiency.

We don't wish to retreat to the countryside and live like the Unabomber in a plywood shack. We believe that people are best off living in cities and cooperating with other like-minded folks. Instead of hoarding ammo and MREs, we're building the skills and forming the conditions and networks that sustain us, our friends, and our neighbors, now and into the future.

Urban homesteading is about preparedness, but we don't like that term very much. It connotes stockpiling things that you hope will keep your ass alive. Survivalism in general is about the fear of death. Urban homesteading is about life—it is a way of life founded on pleasure, not fear. Our preparedness comes not so much through what we have, but what we know. We are recollecting the almost-lost knowledge of our great-grandparents, those most essential of human skill sets: how to tend to plants, how to tend to animals, and how to tend ourselves.

Over the last couple of generations we've given up these skills in exchange for a self-destructive addiction to "convenience," becoming, as a friend of ours likes to say, the only animal that cannot feed itself. We do not make anything anymore, we just consume—we are "consumers," defined solely by our appetites, and empowered only in how we spend a dollar.

We figured it was time to become producers again.

That is what we are trying to do here on our little urban farm: produce food, hack our house to generate power and recycle water, plot revolution, and build community. Changing what and how we eat is at the heart of everything, though. Homegrown food is mind-blowingly fresh and flavorful, 100 percent organic, untainted by disease, blood, or oil, and alive. Trust us, once you discover that lettuce actually has a distinct flavor, or you eat a sweet tomato still warm from the sun, or an orange-yolked egg from your own hen, you will never be satisfied with the pre-packaged and the factory-farmed again. The next step after growing fresh food is using the old home arts to preserve it: pickling, fermenting, drying, and brewing.

Over and over again we've discovered that anything we figure out how to do ourselves tastes better than what the market offers us. If it wasn't, we probably wouldn't keep doing this. Yes, it is a "green" way to live, it is a prepared way to live, it has many virtues, but frankly, it is pleasure that inspires us to do more and more. Get into this a little, and you'll realize that all of your life you've been cheated. Urban homesteading is not about deprivation or suffering, it is about reclaiming your heritage, and your right to real food and real experience.

Make the Shift

We are not alone, and we didn't invent this idea. Urban homesteading is a movement, a quiet movement of sensible people making the smart choice of disconnecting ourselves in healthy ways from an increasingly untenable reality and creating our own culture from the ground up. We live better, we eat better, we're saving the planet. What's not to love?

Anyone can be an urban homesteader, even if you live in an apartment. You can grow more food than you think in a small space: on a balcony, a roof, a side yard. Do you live in a windowless hole? Then use a community garden plot, or claim land and become a pirate gardener. Opportunity abounds even for those of us in the dense metropolitan core. We've met a guy who keeps bees on his roof and harvests hundreds of pounds of honey each year in the middle of Brooklyn.

Most American cities sprawl. They possess tremendous amounts of wasted space. Once you take the red pill and open your eyes, all of that space begs to be cultivated. It is an offense on the level of sin for good land to sit unappreciated and unused under lawn and concrete. The single family dwelling with its defensive swath of front lawn and hidden backyard—the basic unit of the American dream—happens to be the perfect mini-farm. We have a vision of cities greened not by lawns, but by crops, thousands of city gardens collectively forming vast tracts of urban acreage. We each can start with our own patch of land and in so doing inspire others. Since we planted our parkway (that useless space between the sidewalk and the street that is technically city property) with vegetables, several of our neighbors have planted their own victory gardens.

Urban homesteaders are forming organic networks to share knowledge and know-how. What our ancestors took for granted, we have to reinvent. It is hard to figure all this out alone, so we have to help each other. Erik and I have been documenting our homesteading experiences on our own blog, www.homegrownevolution.com. Our posts cover the homesteading basics. Not by ranting, as we have today, but through step-by-step projects and practical advice that will make a homesteader out of you in no time. Our message, as always, is get out there and do something!

≫

Kelly Coyne and Erik Knutzen, authors of *The Urban Homestead,* have become increasingly interested in the concept of urban sustainability since moving to Los Angeles in 1998. In the that time, they've slowly converted their 1920 hilltop bungalow into a mini-farm, and along the way have explored the traditional home arts of baking, pickling, bicycling, and brewing, chronicling all their activities on their blog, Homegrown Evolution: www.homegrownevolution.com.

To Build a Better Future, Start with a Better Question

JEFFREY HOLLENDER

Anyone who contemplates the role of business in society must be capable of simultaneously holding two opposing truths in mind: business is responsible for much of what ails and abets the modern world.

Over the past half century, the astounding rise of the multinational corporation has made business the most powerful force in today's society, surpassing even governments. Just recall FEMA's feeble response to the devastation caused by Hurricane Katrina—Walmart, with its world-class logistical operation and help from scores of nonprofits, proved to be the *real* first responder. Whether it's Walmart providing disaster relief, Genzyme combating disease, Liberty Media holding sway over satellite communications technology, or ConAgra feeding vast swaths of the nation, business has generally proven itself to be the most efficient (though often not always the best) way to get things done.

Business is likewise skilled at creating and reinforcing our experience of life in the industrialized world. Apple designs the way we listen to music; Facebook helps shape our relationships; Rupert Murdoch's News Corp. increasingly frames the world's events and thereby determines more of what we discuss.

But the business world's sheer dominance over human affairs has created a host of unintended consequences, which include the rise of hunger, poverty, and inequality, as well as the depletion of natural resources, global climate change, and the extinction of countless species. While society's ills live outside the balance sheets of the multinational corporations that contribute to these problems, that fact makes them (and us, as consumers of their products and services) no less culpable. So it's more than ironic that the business world's unrivaled power is the very thing that makes for-profit enterprise a critical force for creating a better future. We can't "ease the world's inequities," as Bill Gates once put it, without re-defining the very purpose and possibility of business.

If we're going to harness business to help us build a better future, we've got to learn to ask better questions. No matter what your field of endeavor, the question you ask shapes the answer you get. If you run a company, and you ask, "What can we do to build market share?" you will get a very different answer—and you will create a very different future—than if you ask, "What can we do to build a more sustainable economy?"

For too long, those of us in business have proved adept at posing the first kind of question, but all too inept at considering the second. Here's a question that every business leader should ask, but too few do: "What does the world need most that we are uniquely able to provide?" The question embraces business' vast potential to be a positive force for change. It's a question that forces us to explore how we can develop the new thinking needed to respond to the vexing challenges (and boundless opportunities) that confront the planet.

Corporate responsibility and, more recently, corporate sustainability have been billed as the way forward for businesses committed to thinking beyond the next quarter and taking responsibility for all of their stakeholders. While this is a welcome development, it falls well short of what is needed. Something much deeper and more vital is required, something we at Seventh Generation call "corporate consciousness." We believe sustainability is an important step on the long journey to a regenerative state, where business improves rather than simply maintains our social and environmental support systems.

A company that seeks to rebuild society and restore the environment starts by asking a question like, "How can we change our value proposition from 'selling desirable products' to 'solving difficult problems?'" This mindset helped frame an innovative venture between the French food multinational Groupe Danone (owner of Stonyfield Farm) and Grameen Bank, the pioneering microfinance enterprise. Grameen and Danone have forged a partnership that twines two business objectives that are often in conflict—profitability and responsibility—to create a new kind of hybrid: the "social business." A social business must win customers and compete in the marketplace, but it also seeks to untangle a knotty social problem. In the case of Grameen and Danone's joint venture, the goal is to improve the nutrition of poor children living in Bangladesh.

As first reported in *Fortune* magazine, the two enterprises have combined to build a yogurt factory outside Dhaka. Instead of landing on Danone's bottom line, all the revenue and profits are reinvested in the project. The initiative relies on local Grameen "micro-borrowers" buying cows to produce the necessary milk, which is sold to the factory; on Grameen "micro-vendors," who sell the yogurt door to door; and on Grameen's 6.6 million members, who purchase the ultra-affordable final product for their kids. If the scheme succeeds, Danone will build fifty more factories, for a total capital investment of $25 million.

Some of the dividends from Grameen/Danone won't show up on a traditional balance sheet; but while they're financially intangible, they're still very real. Danone has created a brand-enhancing social dividend for its shareholders and it's expanding into a new market. For its part, Grameen estimates that within eighteen months, two cups of its yogurt per week significantly improves the health of malnourished children. Moreover, this daring venture will provide more proof that capitalism and sustainability can prosperously coexist.

If we can harness the expertise and financial resources of the world's largest companies to address society's problems in a sustainable manner, the future will indeed be brighter. Much of my work, and the work of Seventh Generation, aim to bring about just such a future. Our efforts began several years ago, when we asked ourselves that same, fundamental question: what are we uniquely able to do that the world most needs?

The work we put into answering that question led to the development of our Global Imperatives, which represent the world that we dream of, the world at its very best. These objectives could take as much as a century to achieve—a time horizon that few businesses would tolerate. That's what makes the pursuit of global imperatives so challenging. They require a complicated mix of many efforts: cooperation with other businesses and organizations, our own ongoing education and development, and the need to look systemically at everything we do.

Our foundational imperative is also our first: "As a business, we are committed to being educators and to encouraging those whom we educate to create with us a world of equity and justice, health and well-being."

On the surface, this sounds impossibly romantic and hopelessly naïve. But it doesn't mean that we all join hands and sing "Kumbaya." After all, it's an *imperative*—it calls us to action, and it has catalyzed some very tangible, real-world initiatives. Chief among them is Seventh Generation's collaboration with Oakland, California-based WAGES (Women's Action to Gain Economic Security), a non-profit that creates new jobs and empowers low-income women by organizing and incubating cooperative businesses.

Seventh Generation is committed to helping WAGES create cooperative house-cleaning businesses throughout the United States. The social returns from this venture require investments of capital and business know-how, in the same way that economic returns require them. WAGES provides the initial organizing, training, business systems, support, and nurturing necessary to help the cooperatives become self-sustaining enterprises. But WAGES can't expand these efforts without our assistance. That's why we've paid the salaries of WAGES managers, assisted in training cooperative members on health and safety issues, and used our brand to help promote the Co-ops' services.

Admittedly, WAGES and our half-dozen other social initiatives only begin to push Seventh Generation toward becoming an enterprise that truly restores and enriches society and the environment. But they are a big part of what gets me out of bed every morning and makes me eager to come to work. It's my hope that our example will inspire other companies to move beyond the obligation to be less polluting, less wasteful, "less bad," and to seek out innovative ways to be all nourishing, all replenishing, "all good."

We're just beginning the journey, but this much I know: our collective ability to transform the world requires us to raise our consciousness and ask better questions. What effect on the world do we seek to have? How can we help people realize their full potential? What role are we willing to play? I don't know if these are the "right" questions, but I suspect they'll lead to the right kind of conversations—deliberations that might even push other companies to embrace a model of deeper purpose. Our future depends on it.

❧

Jeffrey Hollender is the cofounder and executive chairman of Seventh Generation, the leader in green household products. He is the co-author, with *Fast Company* founding and senior editor Bill Breen, of *The Responsibility Revolution: How the Next Generation of Businesses Will Win* (San Francisco: Jossey-Bass, 2010), from which portions of this essay are adapted.

Getting Ready
for Change

BILL MOLLISON

I had accepted that we were in real trouble by 1974 (the "Club of Rome" report[1]), and I live as though it is wise to make ready for trouble. After many years as a scientist with the CSIRO Wildlife Survey Section[2] and with the Tasmanian Inland Fisheries Department, I began to protest against the political and industrial systems that I saw were killing us and the world around us. But I soon decided that it was no good persisting with opposition that in the end achieved nothing. I withdrew from society for two years; I did not want to oppose anything ever again and waste my time. I wanted to come back only with something very positive, something that would allow us all to exist without the wholesale collapse of biological systems.

By 1984 it had become clear that many of the systems we had proposed a decade earlier did, in fact, constitute a sustainable earth-care system. Almost all that we had proposed was tested and tried, and where the skills and capital existed, people could make a living from products derived from stable landscapes—although this is not a primary aim of permaculture, which seeks first to stabilize and care for land, then to serve household, regional, and local needs, and only thereafter to produce a surplus for sale or exchange.

The world has changed and will continue to do so. We have two choices with how to move forward: we can sit, curled up in a little ball, rocking back and forth while hoping it will fix itself, or we can get out there (age being no barrier as I'm eighty-two years old) and do something about it.

All of us would acknowledge our own work as modest; it is the totality of such modest work that is impressive. There is so much to do, and there will never be enough people to do it. We must all try to increase our skills, to model trials, and to pass on the results. If a job is not being done, we can form a small group and do it. (When we criticize others, we usually point the finger at ourselves!) It doesn't matter if the work we do carries the "permaculture" label, just that we do it.

If Antarctica sheds all its ice, the sea will rise some seventy meters higher, so I no longer live at sea level but lately ensure a minimum of one hundred meters' elevation. Our beach cabin is for sale!

Even this may bring the coasts close to my door, and I have educated myself about edible seaweeds of this coast. As well, we have now fitted six 10,000-gallon (44,000-liter) tanks to our buildings and outbuildings to catch roof water—enough to provide for our household and gardens in drought.

Over the last ten years, we've planted about two hundred nut pines, sixty chestnuts, five walnuts, twenty apples, twenty pears, thirty plums, and about fifty other fruit trees and vines. All of these could be planted by three people in three months if necessary.

We also have thirty geese, fifty chooks (that's Aussie for chickens), thirty ducks, seven to twenty pigs, and wild rabbits and wallaby, fish and lobsters. The pines, apart from offering us nuts, also provide cones to fuel a Russian stove that heats the house, as well as a cook stove. We are now building a bread oven. Not to forget the ninety to a hundred oaks (to feed geese and pigs) and a few thousand blackwoods (*Acacia melanoxylon*) and other Australian native trees, for honey and fuel and lumber. These are wild-seeding or (in the case of ti-trees) clump-forming.

We have a few "instant" thickly mulched gardens and fifty to a hundred meters of fresh vegetable beds. We invested in a tunnel (un-heated) hothouse for tomatoes, banana, ginger, lemongrass, and so on. This coming year we will invest in solid trellis for grapes, kiwi fruit, hops, and more.

All tank outlet pipes are pressurized to ensure good pressure at the buildings. For wildfire an eight-horsepower diesel pump services underground two-inch lines and hoses. We have life subscriptions to two seed-savers clubs.

HOPE BENEATH OUR FEET

An electric farm system is being designed, and an "X-Trail" vehicle (a four-wheel drive Nissan model) is fitted with a liquid petroleum gas (LPG) tank and lead-free petrol. Along with rainwater storage, we have three dams for native fish, including blackfish and eel.

About twelve months from now, we should have fitted enough solar-electric panels to power all essential functions, putting us well on our way to sell surplus energy to the state power utilities. Building and fitting energy-efficient homes and devices, we should be as well equipped as can be.

I designed and built an energy-efficient house in 1955, and since then have designed about two hundred for clients and friends. There are many people able to do this today, and to retrofit inefficient homes, which in fact I enjoy more than building anew!

My wife Lisa and I are happy, contented, ready to face tough times, and possess the tools to do more. You may get the idea that we are forward planners, and this is how we advise you to behave. None of this is difficult, all of it is beneficial, and quite simply, if you don't do it, you are taking an unwarranted risk, for you, your children, and your society. Change *is* coming, but we have time to make ready.

NOTES
1. www.clubofrome.org/docs/limits.rtf
2. CSIRO: Commonwealth Scientific and Industrial Research Organisation, Australia

❧

A baker, fisherman, CSIRO technical officer, and university-tenured lecturer, Bill Mollison conceived Permaculture and spread its strategies through education. He founded the Permaculture Institute in 1978 to further these goals. He originally self-published his books, articles, and curricula, which entirely funded all Institute work. A passionate teacher, at eighty-two he still conducts courses. By personally planting the seeds of Permaculture in 120 countries, Bill's efforts have been recognized. He is a receiver of the Right Livelihood Award, Outstanding Australian

Achiever Award, Vavilov Medal, Australian Ecology Icon of the Millennium, Steward of Sustainable Agriculture Award, and more. His Web site is www.tagari.com.

Thinking Like an Island

MICHAEL ABLEMAN

The following essay is based on excerpts from a talk given at the 2009 Future of Food conference at Boston University.

I've had a relationship with islands since I was a young man. My attraction has not been to the clichéd tropical fantasy of palm trees and white sand; it is something much deeper—island as metaphor for our existence on earth, representing independence and interdependence, natural limits and boundless space. Island as paradox.

Living on an island as I have for the last ten years, I've realized that you can't get away with anything. The feedback loop is pretty immediate; there is no vast landmass where the ripples of one's actions and interactions can radiate over long distances and large populations in a grand anonymous dilution. If we consider the earth itself as an island floating in a sea of space, in fact "our only ship at sea," as my friend David Brower described it, we might reconsider how we take care of it.

One of the most emblematic places to start, I believe, is with our food. There may be nothing more central to our lives than how we secure our food. Yet the responsibility has been handed over to an industrial system where farms have become factories and food has become a faceless commodity. The results have been disastrous: epidemic levels of childhood obesity and diabetes, polluted groundwater, soil degradation, food that no longer tastes good or is good, and most profound—an almost complete disconnection from the social, cultural, and ecological connections that were once part of agrarian life.

The Polynesian people who first settled the Hawaiian Islands understood how to live within ecological limits. They supported a population of a million people on those islands without any outside inputs. There was a well-prescribed way of living and of managing resources. If someone caught a species of fish out of season, that person was punished. Sounds pretty radical, but they understood something that we have forgotten; that the survival of each one of us is inextricably tied to each other and to the natural world we live in. Those original island peoples knew that the greed of a few could unravel the survival of an entire society.

Hawaii is the most remote populated landmass on earth, and yet it now imports close to 80 percent of its food from the mainland. Food is traveling close to three thousand miles to reach its shores. The costs of this system go well beyond what is being paid at the checkout counter. A place that has such a feeling of wealth and fecundity has become one of the most food insecure places on earth.

When I was sixteen years old, I spent time on another island, the island of Jamaica. My brother and I were taken in by Gretel Hilton and her partner, Uncle Will, who patiently instructed us in some basic survival; how to sharpen a machete and open a coconut, what herbs to use if we were injured or sick, how to cook breadfruit, and how to fish from the cliffs along the sea.

At that time I had no idea that I would eventually devote my life to learning those very skills, writing and teaching about the critical importance of rediscovering our place in Nature and knowing how to grow food. My generation was in the early stages of what is now a total worship of technology; we were invested in our own cleverness, abandoning the intelligence of Nature that had guided humans for thousands of years. We didn't realize the ecological and social price we all would pay for this arrogance.

But Gretel Hilton and Uncle Will were still immersed in the natural history of a place and they were not alone. At that time, most rural families in Jamaica were fairly self sufficient, still had chickens and goats, a breadfruit tree, coconut trees, mango, some cultivated yams and greens and, if they lived near the coast or a river, they fished.

While poverty in the economic sense of the word was endemic then as it is now, this diversity and food quality was a form of national wealth embodied by rural communities that prided themselves on the variety and quality of their fruits, and knowledge, and an intimacy with a place.

The potential for abundance is still there, supported by a tropical climate, rich soils, and plentiful water, but that self-reliance, especially in regards to food, has been replaced by a total dependency on imports from abroad.

Thinking like an island requires that we accept conflict and contradiction; accept that there is nowhere to go, nowhere to run; accept that, as Wes Jackson articulated it, "We are more ignorant than knowledgeable and need to act accordingly."

A farm is like an island; when run well a farm should be self-contained, self-sustaining, its nutrient loop closed and fulfilled from within, not reliant on external inputs. This is farm as ecological system, but in a world where only 1 percent of us are growing the food for the rest, farms have a cultural and educational role as well.

I spent twenty-five years developing a small twelve-and-a half-acre farm and education center in California. Floating in a sea of tract homes and shopping centers, this farm was another type of island. During its heyday, twelve-and-a- half-acres produced one hundred different fruits and vegetables, employed thirty people, provided food for five hundred families, and generated close to a million U.S. dollars in gross income.

Threatened with development, we formed a non-profit organization—the Center for Urban Agriculture—and against all odds raised a million dollars to save that land, placing it under one of the first active agricultural conservation easements in the country.

But the internal struggles of farming in a suburban environment eventually got to me. I found myself longing to live and farm in a place where one's sense of responsibility extended beyond the edge of the lawn. So we moved 1,200 miles north, to an island in British Columbia, Canada.

Boarding the ferry to get to that island was like pulling up the drawbridge. There was this sense that we were leaving the madness of the world behind; Island as refuge. But I soon discovered the great paradox; that those things I thought I was leaving behind were right there as well, and in ways that were more difficult to ignore. Living on an island, I discovered, does not allow for escape, it forces engagement more than disengagement.

We farm on one of the islands original homesteads, 120 acres that sit in the heart of the island's watershed.

One of the most wonderful parts of living on this land are the quiet glimpses into the past that appear in unexpected ways, the sense that we are a part of a long chain of humans on the land, from the native people who first fished its creeks and lake to those who built the original homestead to ourselves, each link informed by the past and by the land itself.

I have always emphasized the importance of land tenure as a critical principle for creating a truly sustainable food system, but now I wonder what land tenure really means. After all, we are just passing through, temporary tenants and caretakers of a larger natural force. All that will ultimately remain is the land, and the best we can do is to leave it more fertile, more alive, and more biologically diverse than we found it, and to use our brief time on the land to feed and to nourish and to inspire.

We are living on the cusp of one of the most significant global changes since the onset of the industrial revolution. We all are acutely aware of climate change, a looming energy crisis, and populations increasing in parts of the world at staggering rates. We know that the most fundamental elements of life, such as soil and water and clean air, are under unprecedented assault. Unlike the Polynesians who had evolved a closed system on their islands, most islands are now wholly dependent on the outside for their most basic needs.

If we are going to be able to move through and survive the massive changes that are taking place in the world, many more of us are going to have to find our way back to the art and craft of growing food.

In October of 2001, I gave a speech to the Bioneers conference in San Francisco in which I proposed that, in memory of the thousands of people who lost their lives at the World Trade Center, a portion of that site be converted to an urban farm, replete with greenhouses and kitchens and an education center. That this farm be established to provide food and jobs year round to those in need and that it became a model of a local agricultural-based economy on the grounds of what used to be a monument to the global economy. My idea was put forth more as metaphor than with any expectation that it would be seriously considered, but a couple of organizations in New York City picked it up and the proposal was officially submitted. The *New York Times* ran a piece on the idea and there were hundred of letters in support, but in the end what was approved was 1,492 stories of cold steel, glass, and concrete, and it was business as usual.

Had that proposal been put forth today, it would have received a more positive reception. Awareness around food—its place in our lives and the precarious nature of the system that brings it to us—has exploded. But while there is an overwhelming embrace of local food and agriculture, there is an enormous chasm between those who eat well and locally (and can afford to do so) and those who cannot. There is a far greater gap between the numbers of eaters who are passionate and enthusiastic and inspired by this movement and the numbers of people whose hands are actually in the soil doing the work.

In the end there is not so much a food crisis or an environmental crisis as there is a crisis in participation. We now have a couple of generations of young people who are not only completely de-natured, they no longer know how to use their hands for anything other than pushing keys on a keyboard. The revolution may be talked about online but it cannot take place online.

In 1989, the island of Cuba faced mass starvation, the result of having lost access to food and agricultural supplies from the former Soviet Union. Cuba responded by creating a world-class model of urban and rural agriculture based on low inputs. The Cubans did not green their agricultural system because it was the right thing to do, they did it because they had to.

I don't believe that the kind of major structural change that will be required to turn things around in industrialized countries will happen until it has to either, until the impacts become personal. But the hopeful part is that humans have an incredible capacity for compassion, ingenuity, creativity, and resourcefulness; qualities that come out especially under duress and crisis. There are many historical examples of this when normal day-to-day reality is suspended; Cuba is but one.

I used to say chefs had received almost mythical rock 'n' roll status and that it was time for farmers to receive that same attention. But the real shift we need cannot take place when only 1 percent of us are doing the work to grow the food for the rest, while everyone else is cheering us on. I love the attention, but farming is not a spectator sport.

So I've been telling folks to "make friends with a farmer, you're going to need them." For I am certain that as the current global industrial experiment continues to unravel, agriculture will once again return to it's rightful place; to the heart, the center of our society.

So those of us who are re-educating ourselves, re-discovering our place in Nature, must work to refine our skills and diligently work to create the local and regional models. For I am sure that the day will come when we will be sought after, looked to for leadership and guidance, when our farms will be the living models, the repositories that kept this sacred and essential knowledge alive.

Thinking like an island, imagining our world floating like a ship in an infinite sea of space, its soils and water and atmosphere finely tuned, carefully balanced to support life, provides us with some sense of boundaries, and hopefully the humility to recognize the fragile nature of our existence.

❧

Michael Ableman is a farmer, educator, founder, and executive director emeritus of the Center for Urban Agriculture, where he farmed from 1981 to 2001. He is the author and photographer of *From the Good Earth*

(Abrams, 1993), *On Good Land* (Chronicle Books, 1998), and *Fields of Plenty* (Chronicle Books, 2005) and is the subject of the award-winning PBS film *Beyond Organic,* narrated by Meryl Streep. For more information, visit www.fieldsofplenty.com and www.foxglove farmbc.ca.

The World Is Falling Apart!
What Should I Do?

SUSAN BARTLETT

I have the honor of being a state senator in the small but distinctive state of Vermont. I am the chair of the senate appropriations committee, so my job is to lead a committee in making difficult decisions, sometimes seemingly overwhelming decisions. I find that these skills apply very well to the rest of my life and the world that we all live in today.

There is no doubt that our world is facing severe environmental stress, many folks believe that it is an environmental catastrophe and we are doomed. The polar ice is melting, there is too much rain in some places and other places are turning into deserts. The cost of oil is out of sight and filling up your car is a horrible experience. Food is more and more expensive, college will never be affordable for my kids and who can afford health care, don't bother me with this environmental stuff, who cares?

Our lives are so busy it all just seems too much to cope with. So many of us just keep right on doing the same stuff we have always done; or as I like to say, "We've cut the board three times and it's still too short!" That's where I say, the world is falling apart and what should I do?

As I see it any individual has two choices; stick your head in the sand or look around and see what reasonable things you can do to help make a difference. If you chose to stick your head in the sand, well you are leaving other parts of your body quite vulnerable, so I recommend against sticking your head in the sand.

My solution to the world falling apart was to run for public office. I was fed up with politicians who didn't represent my priorities and the priorities of my friends. The current senator was retiring and everyone knew who the next senator would be, another guy who didn't represent my values. It wasn't sitting very well with me. I believed that voters deserve a choice and I was tired of not having a choice.

One night as my husband and I were in bed reading, I turned to him and said, "I have an idea, I want to run for the state senate." Bill, who is a very calm, organized, thoughtful man, and is used to my hair-brained ideas, put his book down, looked at me for a moment, and said, "That's the best idea you've had in a long time." Thus was launched my political career.

We had no idea how to run a campaign and there was no assistance from the state party. There was a handbook that had been written to help candidates run for house seats. Bill and I took the book, followed the directions, and shocked the entire state when I won. I ran on ideas that were important to me and I felt were important to my region of the state. I spoke from my heart. I wasn't afraid to say I didn't have all the answers, but that I could promise that I would always listen to folks, be respectful and then do what I thought was best for all of us.

Today Americans have little respect for politicians. Whether you like it or not, politicians represent our society. If you dismiss politics as not important or too complicated or a nasty business, just remember it's your responsibility to place folks in office who represent what you feel is important for your life and your world. No one can represent your values better than you can. Go ahead, take a risk, get directly involved in politics.

If you don't like public policy, you can change it. That means on your local school boards (how about local foods and good foods in your schools?), local planning commissions (how about green zones and protecting wetlands?), local government (where are those bike paths?), or state office (we want clean water and clean air and diversified sources of power.) You have the ability to change public policy.

I know, it's a scary thought, getting involved in politics. What do you know? What do you mean "get up in public and speak"? You don't have to have all the answers, find your issue, find your passion and go for it. If you fail in that responsibility to get involved, well you have no one to blame for the failure of public policy except yourself. If you can't become a candidate, then actively support a person who is willing to take on the job. Democracy is a wonderful institution if we use it. When we fail to participate, we are giving away one of the most precious things in the world: our right to shape our future.

What I'm trying to say is that we all can chose to be overwhelmed, or we can choose to take simple steps. When I decided to run for office, it was a remarkably simple step. It just seemed the obvious thing to do. I wanted a choice and I wanted my friends to have a choice.

Since I first wrote this article I have decided to follow my own advice. I am running for Governor of the state of Vermont. I need someone who represents the values of my friends and myself. One never knows where simple steps will lead us, but a least we know we are in motion. It's really that simple.

⤳

Senator Susan Bartlett lives in Vermont with her husband Bill, Lulu the labradoodle, twin cats Howard and Dean, and two Morgan horses. When she's not involved in the political world, Susan loves to read, travel, cook, and take agility classes with Lulu.

Challenging a Corporation to Clean Up Its Act

THAÏS MAZUR

Protecting the wild core of our sustenance requires vigilance and some-times dedicated action. When I gave birth to my only child, I chose to leave the crowded city and move to a five-acre homestead on the rugged coastline of Northern California. As a new mother, I envisioned a healthy, pristine environment in which to teach my daughter the joys of living close to the earth. In a totally unexpected twist of fate, my vision quickly shifted.

One day, while driving to town, I stopped at a red light. My year-old daughter babbled to herself in the back seat. As I sat lost in thought, my eyes drifted to the west where I saw the familiar cyclone fence cordon-ing off the Georgia Pacific Corporation (GP) lumber mill that had recently shut down. The mill site sits three miles north of my home and covers 434 acres, stretching along four miles of coastline and making up one-third of the town of Fort Bragg. It was used as a military base, an Indian reservation, and successive lumber mill sites for close to 150 years. The longstanding timber industry practices had turned a breathtaking head-lands teeming with life into a wasteland of barren soil and abandoned buildings.

As I pondered the history of the land, I felt, more than heard, a call-ing that shook me to my bones. It started as a low moan and grew to a high-pitched wail. My chest felt as though it would burst from tremen-dous grief. The sharp blast of a car horn snapped me back to my place behind the wheel. The light had turned green and cars were impatiently maneuvering around me.

Weeks later, I learned that the Georgia Pacific property was put up for sale. I knew Georgia Pacific was the largest landholder in the United

States and one of the five largest corporations. Their financial success, based on extraction, was a perfect fit for the harvesting of California's productive timberlands, one of our state's largest economic resources.

The headlands on which the now-defunct mill sat had been off limits to local citizens for more than a century, making Fort Bragg a coastal town without a coast. Public access was quickly in the minds of local residents when news first broke of the mill's last closure. GP had finally depleted the timber resource. There was a problem, however. The land was rumored to need a great deal of environmental remediation.

My good friend Loie Rosenkrantz, my husband, and I began ruminating on the good and not so good things that could happen in place of a mill.

At a satellite Bioneers conference held in Caspar, a small town south of Fort Bragg, we listened as John Todd, a pioneer in bioremediation, gave a lecture on his work using plants to clean polluted lakes and streams. Conference participants gathered to discuss possibilities related to the old mill. The three of us decided that we would invite our community to bring forth visions for the former mill site.

Fort Bragg numbers just over six thousand residents and sits directly east of the mill site with coastal Highway 1 in between. Under the name of North Coast Action, we organized and led community gatherings to brainstorm the future of the mill site.

We upheld the values of transparency and participatory democracy, offering a place at the table for everyone. People from all walks of life, including fishermen, timber harvesters, and real estate agents, attended the meetings. Everything was proposed, from a marine research center to a performing arts center to a sustainable mill. Many coastal residents wanted to see a museum dedicated to a small group of Pomo Indians still residing on the land. One night, after a particularly energized gathering, a man approached me wearing a worn jacket and faded workman's cap. He pulled me aside and in a hushed voice told me that he had worked as a millwright and was aware that the mill property was badly polluted with highly toxic PCBs (polychlorinated biphenyls) and other chemicals. He didn't want to reveal his name for fear of retribution by his past employer. The next week, it seemed that everywhere I went, I was hearing story after story about toxins on the mill site. Loie was also hearing stories about PCB spills, burying of toxic waste, and burning of toxic materials in the mill's powerhouse.

With guidance from a friend who is a veteran toxicologist, we placed an ad in our local newspaper asking anyone aware of any toxins buried, burned, or dumped on the mill site to contact us. By telephone and personal interviews, people anonymously recounted stories of dumping PCB transformers into the ocean and mill ponds, throwing buckets of PCB oil onto the soil, burning toxic waste in the incinerator, and burning open-air piles of toxic materials on the rim of the headlands. Ditches and pipes leading from mill buildings carried waste products and solvents directly out to the beaches and into the ocean.

The information was overwhelming and disheartening. If all the things mentioned had happened, and more that we hadn't been told about, then we had a deadly situation on our hands and a fight that would demand a tremendous commitment if we were to heal the land and restore health in our community. I became more determined than ever to uncover the truth.

My connection with the land, my family, and the community is the foundation for my life, my love, and my work in the world, both as an artist and activist. This love can fill my heart with joy and make me fierce when something I love is threatened. I attempt to teach my daughter the power of this fierceness by taking action. The year I was pregnant several of my women friends died of cancer. Each death heightened my awareness of the toxic state of our environment. I feared for my daughter's health in the future. Now, here I was, living downwind from a defunct lumber mill with a century-old toxic legacy!

Loie, my husband, and I began attending city council meetings, expressing our concern about the toxins on the mill site. We were assured again and again by corporate spokespeople that the land was "no more polluted than a gas station."

However, the more we delved into the issue, the more we realized that this statement was simply a lie. The land was vastly poisoned, and we felt strongly that the corporation that poisoned it should clean it up—not cover it up. We began to meet once a week to plan our strategy, and those meetings quickly grew to three or four times a week. We worked long, hard, furious hours, learning about toxins, reading eight-inch-thick reports, and talking to scientists. I had worked as an investigative journalist for more than a decade and was adept at digesting vast amounts of information. Even still, the learning curve was steep. We continued voicing our concerns to the city council and asking that the city staff take the toxic nature of the mill site seriously. We worked to educate our community and forge relationships with allies beyond.

We tried to educate and work collaboratively with the local city government, but it became obvious to us that our community had lived far too long under the shadow of corporate strong-arming. The hypnotic pull of the old mill-town history created a denial that was pervasive. It wasn't long before local politicians shunned us, and threats were made on our lives. Nevertheless, we continued to educate, organize, and speak out.

After four years of working to uncover this toxic legacy we discovered that high levels of dioxin—the most deadly toxin known to humans, second only to radiation—had been discovered in the mill site's soil. Rumor had it that the dioxin had been created by burning treated redwood bark in the mill's powerhouse, which generated huge amounts of fly ash. People began to come forward with stories of Georgia Pacific transporting fly ash off the property and dumping it on school grounds, ranches, and in the gardens of residents. What we thought had been contained to the mill site was contaminating soil up and down the coast.

Our scientist friends began referring to the coastline as "Behind the Redwood Curtain." The testing could take years and the cleanup could be immense!

Stories of people on the coast with rare forms of cancer, brain tumors, and autoimmune disorders began to surface. People talked about alarming numbers of miscarriages, birth defects, and infant mortality when the mill was in operation. Dioxins have been shown to accumulate in humans and wildlife due to their lipophilic properties, and they are known as mutagens (chemical agents that change genetic information) and suspected human carcinogens. Generations to come would inherit GP's toxic legacy unless the mill site was cleaned up.

Shortly after we found out about the dioxin, Georgia Pacific Corporation was purchased by Koch Industries, the largest privately held company in the world. Owned by the Koch brothers, Koch Industries holds a less than stellar environmental record. In 2000, Koch was indicted on ninety-seven counts of violating federal clean air and hazardous waste laws. The charges came less than one year after the company was slapped with the largest civil penalty ever levied under federal environmental statutes. That same year, Koch Industries was ordered by the Justice Department and the California EPA to pay the largest civil fine ever imposed on a company under any federal environmental law to resolve claims related to more than three hundred oil spills from its pipelines and oil facilities in six states.

Suddenly, the stakes were higher. Shortly after Koch took possession of Georgia Pacific Corporation, they opened the gates in the month of July for the community to view fireworks from the mill site. This was after the tests had reported dioxins on the property, with the findings well publicized in the local media. Women from the community banded together and made up signs warning people of the potential danger to human life on the mill site. They stood outside the gates carrying the message in English and Spanish: "Keep Out—Danger, Dioxins!" Infuriated by the corporate scheme to tell the public the site was safe while blatantly putting us in potential danger of life-threatening toxins, we knew it was time to recruit a hard-hitting environmental agency.

The current oversight agency for the mill site, chosen by Georgia Pacific Corporation, was the Regional Water Quality Control Board. For the past eighteen years, this state agency, through a gentlemen's agreement, had overseen all the lumber counties in Northern California, from the Bay Area to the Oregon border. That was good insofar as it went, but for sites like ours, it was insufficient for a full-site characterization of contaminated soil.

We decided, with the help of many community members, to recruit the Department of Toxic Substances Control (DTSC), an arm of the

California EPA. DTSC is known for enforcing stringent regulations, especially on corporate polluters. Many people warned that we could never make this happen now that Koch was part of the mix. We were told that it was not only impossible, but that it was dangerous. We took a deep breath.

Methodically, step by step, letter after letter, meeting after meeting, month after month, we worked toward our goal. We educated the public about toxins to garner support. We wrote countless Letters to the Editor in our local newspaper. We were interviewed on weekly radio programs. Threats against us increased and so did our fear of being victims of retribution.

One year later, exhausted but not defeated, we made it happen. DTSC took over as the lead oversight agency. We were relieved, and at the same time aware that our work was far from finished.

Today our continuous attention is needed to help ensure that a thorough clean-up takes place. It took a decade to get here. We have successfully engaged Georgia Pacific and Koch Industries to use bioremediation and mycoremediation—a process using mushrooms to uptake toxins from soil. People faced with similar issues across the country call us, wanting to know how we accomplished this seemingly impossible task.

I know this story is but a microcosm of the challenge our world is currently facing. I also know that a small handful of people, through a shared vision and commitment, created an important change.

This place where I have chosen to live and raise my child reflects the boundless love I carry in my role as a mother and citizen of this planet. I must not only protect and nurture my daughter; I must include the community she is growing up in, the air she breathes, the food she eats, and the water she drinks. On a global scale, the same is true if we are to survive.

We currently live in one of the most challenging times in our planet's history with the rapid destruction of the environment and daily extermination of entire species of plants and animals. Many people today are disillusioned and looking for an alternative to the dead-end system that is responsible for the ongoing desecration. Under the shadow of corporate greed, the price of human suffering and a poisoned world is causing our quality of life to deteriorate rapidly.

No ordinary human would have the inclination to spend his or her life working endlessly toward solving the problems of the world. There is no money to be made, no fame or power. And yet, many people around the globe are standing up to save the planet, deeply trusting what they know to be right.

The world is in crisis and the problems to be solved are endless. We don't need to take them all on, but ultimately we need to stay awake and remember that whatever befalls this planet affects us all. Conscious thought, small actions, large actions—all of these count toward redirecting the destructive course we are on. There is no guarantee that we will be successful, but that is not the point. The point is that the only way things are going to change is through us making it happen.

❧

Thaïs Mazur is an award-winning filmmaker, journalist, and author of *Warrior Mothers: Stories to Awaken the Flames of the Heart.* She is renowned for her in-depth reporting, including a four-part story on Aborigines and uranium mining and breaking the story on the Three Mile Island nuclear accident. She is a founding member of North Coast Action and is one of the leaders promoting mycoremediation to cleanup toxins on the Georgia Pacific Mill Site in Fort Bragg, California. Currently, she is producing and directing a documentary about mycoremediation on the Mendocino coast of California. The film trailer can be seen at www.earthbeatmedia.org.

Nothing Else Matters

DERRICK JENSEN

We need to stop this culture that is killing the planet. Nothing else matters. NOTHING else matters. The only measure by which we will be judged by the people who come after is the health of the land base, because that is what is going to support them. They're not going to care about how hard we tried. They're not going to care about whether we were nice people. They're not going to care about whether we were non-violent or violent. They're not going to care about whether we grieved the murder of the planet. They're not going to care about whether we were enlightened or not enlightened. They're not going to care about what sort of excuses we have to not act ("I'm too stressed to think about it," or, "It's too big and scary," or "I'm too busy," or "But those in power will kill us if we effectively act against them," or "If we fight back we run the risk of becoming like they are," or any of a thousand other excuses we've all heard too many times). They're not going to care how simply we lived. They're not going to care about how pure we were in thought or action. They're not going to care if we became the change we wished to see. They're not going to care about whether we voted Democrat, Republican, Green, Libertarian, or not at all. They're not going to care if we wrote really great books about it. They're not going to care about whether we had "compassion" for the CEOs and politicians who are running this deathly economy. They are going to care about whether they can breathe the air and drink the water. They are going to care about whether the land can support them. We can fantasize all we want about some groovy eco-socialist utopia, and if the people can't breathe the air, it doesn't fucking matter. Nothing fucking matters but that we stop this culture from killing the planet.

It's embarrassing even to have to say this. The land is the source of everything. If you have no planet, you have no economic system. If you have no planet, you have no spirituality. If you have no planet, you can't ask this question. If you have no planet, nobody can ask any questions.

I want people to get over their precious little narcissistic selves, and actually work to protect the land where they live. And one of the good things about everything being so fucked up is that no matter where you look there is good work to be done. We need it all. We need people to take out dams, and we need people to knock out electrical infrastructures. We need people to protest and to chain themselves to trees. We also need people working to ensure that as many people as possible are equipped to deal with the fall-out when the collapse comes. We need people working to teach others what wild plants to eat, what plants are natural antibiotics. We need people teaching others how to purify water, how to build shelters. All of this can look like supporting traditional, local knowledge, it can look like starting neighborhood gardens, it can look like planting local varieties of medicinal herbs, and it can look like teaching people how to sing.

I want people to work to stop the destruction. No! I don't want people to *work* to stop the destruction. I want them to stop it. There's all the difference in the world between those statements. Failure is not an option. I think about a line by Søren Kierkegaard: "Twaddle, rubbish, and gossip is what people want, not action. . . . The secret of life is to chatter freely about all one wishes to do and how one is always being prevented—-and then do nothing." Well, the time is long since past for that.

What I really want is for people to think for themselves and feel for themselves and to listen to their own land base and to ask that land base, "What must we do?" I want people to start a relationship with the land where they live. Ask that land what it needs from them. Because, once again, the land is the basis for everything.

The dominant culture is psychopathological. It kills every indigenous culture it encounters. It kills all that is wild. It is killing the planet. There is nowhere we can run. It is time to fight back.

People aren't going to care *how* we live, individually or collectively. They will care about whether we stop this culture from killing the planet, so that they may live. They will care about how far destroyed is the world that we leave to them.

What I want is for people to find what they love, and defend their beloved.

⌁

Hailed as the philosopher-poet of the environmental movement, Derrick Jensen is author of fifteen books, including *Endgame* and *A Language Older Than Words*. He holds a degree in creative writing from Eastern Washington University and a degree in mineral engineering physics from the Colorado School of Mines, and has taught at Eastern Washington University and Pelican Bay State Prison. He has packed university auditoriums, conferences, and bookstores across the nation, stirring them with revolutionary spirit. His Web site is www.derrickjensen.org.

CHAPTER FIVE

The Body of the World

Eyes Wide Open

CHAMELI GAD ARDAGH

If we could see the miracle of a single flower clearly, our whole life would change.
–Buddha

In some indigenous cultures a core practice is to visit the same place in nature every day and to keep discovering and exploring the changing and shifting details of the place. This kind of practice radically opens our sensory receptivity so that we can feel all the movements around us as our own movements, and our own actions will therefore arise from an expanded sense of self. It is as if our eyes truly open for the first time and from this clear perspective the universe almost becomes technicolored; everything seems fluid and pulsating with life. Many of us remember this kind of receptivity from our childhood, when every little stone and stick were portals into exciting adventures, and we had not yet been caught in what Albert Einstein calls "the optical illusion of the human mind"; the feeling of separation from the earth and all her beings.

When we take time to connect with our earth from a more silent place within, to look and to listen with soft receptivity, we discover a world full of vibrant life. It seems as if the world has revealed itself from its hiding place behind a two-dimensional gray blanket, but the change has of course occurred within us. Connected to ourselves on a level deeper than just the busy activity of our thoughts, we can see clearly again. Normally we live mostly inside the little world of our own thoughts, and when we look at nature as we would look at a picture, at first it doesn't seem like much is going on. But as we open our senses, more and more of the wonder and the interconnectedness of it all is revealed to us.

It is humbling to become aware of the two-way exchange that takes place when we start to open to nature as a teacher. We sometimes set out on a mission thinking that we are saving the environment, and in the process we may discover that on this planet human beings are babies compared to most of the other species around us. Of course, we do what we can to respect and care for nature, but perhaps a big shift will come when we realize that maybe it is nature that will save us.

This radical shift of perspective also happened for the women in Kenya who took part in the Green Belt Movement, when they discovered that when they cared for the trees, the trees cared for them back in surprising ways.

In December 2004 a woman wearing brightly colored clothes emphasizing the deep tone of her skin was standing at the podium of the Nobel Institute in Oslo to receive the Nobel Peace Prize. The fact that she was the first African woman to receive the prize was a sign of a radical change in women's position in the world. That she received the prize for planting trees expanded our understanding of peace and honored our interdependency with Mother Earth. Yes, Wangari Maathai received the Nobel Peace Price for planting trees.

Thirty years ago, in Wangari Maathai's home country of Kenya, 90 percent of the forest had been chopped down, transforming the land into a desert. The job of collecting firewood for meal preparation belonged to the women and girls, and they had to spend hours away from home looking for the few branches that were to be found. Wangari watched all this and decided that there must be a way to take better care of the land and of the women and girls. So she planted a tree. And then she planted another. Soon she realized that it would take an awful lot of time to cover the land with trees if she were to plant them all by herself. So she started to pay women and girls a small amount for each sapling they grew, to help her plant trees all over the country. A movement took hold. It was called the Green Belt Movement, and with each passing year, more and more land was blessed with trees.

But something else was growing as the women planted those trees. Not only the trees were taking root. The women and the girls began to see that they could change their land for the benefit of everyone. They saw that they were making a difference.

They looked at themselves and at each other with a new sense of respect and confidence. They started to trust their own wisdom, recognizing it to be just as important as that of the men. They saw that they deserved to be treated with equality and dignity. But of course these changes were not well received by everybody. The president arrested Wangari many times. For almost thirty years, she was ridiculed and even physically threatened for planting ideas of equality and democracy in people's heads, especially in women's. But she never gave up. She found the strength to continue, along with thousands of women and girls, who were blossoming right along with the trees. Thirty million trees were planted in Africa, one tree at a time, transforming landscapes—both the external one and the internal one of the people. Women planted trees, and in return the trees planted wisdom, hope, and strength in their hearts.

> The woman is planting a tree in the world
> On her knees, like someone in prayer,
> Among the remains of the many trees
> That the storm has broken down.
> She must try again, perhaps one at last
> Will be left to grow in peace.
> –Halldis Moren Vesaas

❧

Chameli Ardagh is one of the world's leading pioneers on contemporary feminine spirituality. She is the director, founder, and senior teacher with the Awakening Women Institute. She is the initiator of a global network of women's groups, a sought-after workshop leader, the author of two books and a series of DVDs on feminine empowerment and spirituality, a trained actress, and a psychotherapist. She delivers public talks, seminars, and trainings to women internationally and is dedicated to assisting women to awaken and to live more empowered and joy-filled lives. Visit www.awakeningwomen.com for more information.

The Healing Power
of Nature

DIANE ACKERMAN

When summer blows through the willows, I love to ramble in an open field near my house, where Queen Anne's lace flutters like doilies beside purple coneflowers. Although I've never harvested the carrot-like roots of Queen Anne's lace, I have taken essence of coneflower (*echinacea*) as a tonic to keep colds at bay. Many people practice such homeopathy—swallowing minute amounts of herbs as curatives for an assortment of ills—and in a sense that's what most of us do, psychologically, when we go out into nature. We drink briefly from its miracle waters. We inoculate ourselves against the aridity of a routine workaday life.

Wild is what we call it, a word tottering between fear and praise. Wild ideas are alluring, impulsive, unpredictable, ideas with wings and hooves. Being with wild animals—whether they're squirrels in the backyard, or heavily antlered elk in Yellowstone—reminds us of our own wildness, thrills the animal part of us that loves the feel of sunlight and the succulence of fresh water, is alert to danger and soothed by the familiar sounds of family and herd. It's sad that we don't respect the struggles and talents of other animals, but I'm more concerned about the price we pay for that haughtiness. We've evolved to live tribally in a kingdom of neighbors, human neighbors and animal neighbors.

When we spend most of our lives indoors, what becomes of our own wilderness? Safe and dry in our homes, clean and well lit, at arm's length from the weedy chaos outside, no longer prey to weather and wild, we can lose our inner compass. A few years ago, for instance, I broke a bone in my foot. And it took a year and a half, four months in a wheelchair, and, finally, a bone graft and titanium screw to heal. For an active person, being so helpless and limited is a nightmare. But the hardest thing about that injury was how it separated me from nature, whose green anthem stirs me, whose moods fascinate me, whose rocks and birds help define my sense of belonging. Even if I'm feeling low, I can always find solace in nature, a restorative when dealing with pain. Wonder heals through an alchemy of mind. But, exiled from Paradise, where could I turn? Once knitted into nature, I felt myself slowly unraveling. Standing upright may be our hallmark, and a towering success, but sometimes bone, joint, and spine can't live up to the challenge and act subversive. A house of bones, the Elizabethans called the body. Imprisoned by my need to heal, I craved the outdoors. To heal I needed to rest, lie low, shelve things, restrict myself, be willing to sacrifice pleasure for recovery. But I only managed it with grace when I rented an electric scooter, climbed aboard, and crept out into the sunlight and among the birds and trees for an hour or so each day. I also had friends drive me out into the country. Those doses of sunlight and wildlife were my salvation. Even a small park or yard can be wilderness enough.

When I'm in a rainforest I caress it with all my senses and am grateful for the privilege, yet I also love temperate forests, scrublands, lakeshores, glaciers, even city parks. One doesn't have to leave home to encounter the exotic. Our human habitat encompasses rolling veldts and mown lawns, remote deserts and the greater wilderness of cities—all "natural" ecosystems. Many animals inhabit the small patch of woods in my backyard, for example, from deer, raccoons, skunks, wild turkeys, garter snakes, and other large fauna down to spiders, moths, and swarming insects. I spend happy hours there watching the natural world bustle about its business. The animals all seem busy, feeding themselves and their families, running one urgent errand or another. Their behaviors remind us of our own, their triumphs teaching us about the indomitableness of life.

I've had the privilege of traveling the world to behold some fascinating animals and landscapes, but I know that one doesn't need to go to the ends of the earth to find an abundance of life, or to feel connected to nature. I felt rapture recently while riding a bike along a country road just as a red-tailed hawk flew very low overhead, showing me the brown-

and-white speckled bloomers of its legs and a bright red tail through which the sun shone as if through stained glass. We're lucky to be alive at a time when whales still swim in the oceans, and hawks still fly through the skies. Alas, one day, through our negligence, they may be gone.

There are noble reasons for protecting the environment—one might argue that it's our moral duty, as good citizens of the planet, not to destroy its natural wonders. There are also mercenary reasons: the vanishing rain-forests contain pharmaceuticals we might need to survive; the Antarctic icecap holds a vast store of fresh drinking water; thick forests ensure that we'll have oxygen to breathe. But another reason is older and less tangible, a matter of ecopsychology. We need a healthy, thriving, bustling natural world so that *we* can be healthy, so that *we* can feel whole. Our word "whole" comes from the same ancient root as "holy." It was one of the first concepts that human beings needed to express, and it meant the healthy interrelatedness of all things (and the miracle of it all). "Mother Earth," we often call the planet. If earth *is* our mother, then we have many siblings among the other animals, many rooms in our home. Most of the time we forget that simple truth, and even pretend that we could live outside nature, that nature doesn't include us.

We really are terribly confused about our relationship with nature. On the one hand, we like to live in houses that are tidy and clean, and if nature should be rude enough to enter—in the form of a bat in the attic, or a mouse in the kitchen, or a cockroach crawling along the skirting boards—we stalk it with the blood-lust of a tabby-cat; we resort to chemical warfare. We don't even like dust around us. In fact, we judge people harshly if their house is full of dust and dirt. And yet, on the other hand, we just as obsessively bring nature indoors. We can touch a light switch and daylight floods the room. We can turn a dial and suddenly it feels like summer or winter. We live in a perpetual breeze or bake of our devising. We scent everything that touches our lives. We fill our houses with flowers. We try hard to remove ourselves from all the dramas and sensations of nature, and yet without them we feel lost and disconnected. So, subconsciously, we bring them right back indoors again.

Then we obsessively visit nature—we go swimming, jogging, or cross-country skiing, we take strolls in a park. Confusing, isn't it?

For the most part, when we go to psychologists we don't discuss how divorced we feel from nature, how destructive that can be, or the tonic value of reacquainting ourselves with nature's charms, the charms we fell in love with when we were children, when nature was a kingdom of wonder, play, self-discovery, and freedom. A special loneliness comes from exiling ourselves from nature. But even my saying that will strike many people as a romantic affectation. After all, we are civilized now, we don't play by nature's rules anymore, we control our own destiny, we don't need nature, right? That attitude is so deeply ingrained in our modern culture that most people assume it's a given and don't worry about nature when they consider improving the important relationships in their lives. It's a tragic oversight, but I can understand why that attitude is so appealing. Nature is crude and erotic, chaotic and profuse, rampant and zealous, brutal and violent, uncontrollable despite our best efforts, and completely uninhibited. Small wonder the natural world terrifies many people and also embarrasses the prim Puritans among us. But most people find nature restorative, cleansing, nourishing in a deeply personal way.

Sometimes it's hard for us collectors of such rarities as paintings, buttons, china, or fossils to understand that we ourselves are rare. We are unique life forms not because of our numbers, but because of the unlikeliness of our being here at all, the pace of our evolution, our powerful grip on the whole planet, and the precariousness of our future. We are evolutionary whiz-kids who are better able to transform the world than to understand it. Other animals cannot evolve fast enough to cope with us. If we destroy their future, we may lose our own. But because vast herds of humans dwell on the planet, we assume that we are invulnerable. Because our cunning has allowed us to harness great rivers, and fly through the sky, and even add our artifacts to the sum of creation, we assume we are omnipotent. Because we have invented an arbitrary way to frame the doings of nature, which we call "time," we assume our species will last forever. But that may not be true. There are little-known species alive among us right now, which have lived on the planet for millions of years longer than we have, but which may perish without our even noticing,

In my career, I've had the luck of going to some exotic and astonishing landscapes in pursuit of the marvelous—from the Amazon to the Antarctic—but fortunately the marvelous is a weed species. It grows everywhere, in backyards and in ditches. Sometimes we need to be taught how and where to find it, but it's always there, waiting, full of magic.

Zoologist Karl von Frisch once described his study of the honeybee (which he adored) as a magic well that replenishes itself endlessly. The same is true for any facet of nature. However much water you draw from it, you always find more waiting for you. It is summer in North America. The well of nature is full today. Time to go outside and take a drink.

❧

Poet, essayist, and naturalist Diane Ackerman is the author of two dozen highly acclaimed works of nonfiction and poetry, including *A Natural History of the Senses*—a book beloved by millions of readers all over the world. Humans might luxuriate in the idea of being "in" nature, but Diane has taught generations that we *are* nature—for "no facet of nature is as unlikely as we, the tiny bipeds with the giant dreams." Visit her online at www.dianeackerman.com.

A Sense of Place—
A Sense of Self

IAN McCALLUM

"Know thyself."
  –Apollo

In 1973 the biologist Theodosius Dobzhansky wrote "Nothing in biology makes sense except in the light of evolution." I would like to suggest, in that same light, that nothing in psychology makes sense either. To me, it will be almost impossible to know ourselves without an understanding and appreciation of what and where we have come from and of how we have survived as a species. Has it ever occurred to you that we are living evidence of a little over two million years of hominin survival, that we share a common one-hundred-million-year bloodline with every mammal and that our ancestors are in our genes? All mammals—from bats and polecats to tigers and chimpanzees—share more than 90 percent of our DNA. Crocodiles and birds share more than 80 percent; fruit flies around 40 percent. A fungus is closer to being human (22 percent of the human genome) than a plant (10–15 percent). We are not only related, we belong. The animals are in our blood. But let us not forget that we are in their blood also—we too gnash and gnaw. We too have our alarm calls, our cries of territory, of sexual display and discovery. We experience fear, frustration, and rage, and we are not the only ones who die of a broken spirit. Let's not forget that the landscape is in our skin as well—every element from silica and hydrogen to lithium, phosphorous and gold can be traced in the human body. The poets are right. We are the dust of the earth and of the stars.

I have a notion that our sense of self, our sense of who we are in the world and of where we belong, is intimately associated with a deep historical sense of landscape—an ancient memory of origins, of where we have come from and of the shared survival strategies of all living things. For some, this landscape is the desert and the open plains and everything associated with it—not just the sight of them, but the sound, the feel, the smell, the taste, the movement. For others it is the mountains and then there are those whose sense of self is intimately linked to ice and water. "This place is in my blood," we sometimes say of these places or, more poignantly, "I feel as if I have come home." To lose this sense of connection with the landscape is to suffer one of the most overlooked psychiatric disorders of our time. It is a condition that I call ecological amnesia. We have forgotten our wild heritage, of where we have come from and of who we are—the human animal. In his poem "The Panther," Rainer Maria Rilke describes this amnesia brilliantly. Putting himself into the skin of a panther that has spent its life behind the bars of a cage, he reenacts an image of wildness that you and I can readily relate to, but which has long since been "barred" from our lives:

> Only at times, the curtain of the pupils
> lifts, quietly. An image enters in,
> rushes down through the tensed, arrested muscles,
> plunges into the heart and is gone.

Characterized by a restless sense of displacement, this amnesia presents as a kind of homesickness, identity confusion, and a low-grade depression in which landscape, animals, geography, and a spirit of place are either completely absent from the vocabulary of the patient, or, when asked about, these very *conditions* of life are described as though they were mere *aspects* of it. When reminded, these patients are often caught by surprise by the emotional impact of the questions put to them: "When you were a child, did you have a tree in your garden?" "Were there mountains near your town?" "Was there a stream near your home that you used to play in and do you know if it's still flowing?" "Do you miss it?"

"How old were you when you first saw the ocean?" "Did you have pet animals?" "What is your favorite wild animal and why?" "What is your favorite landscape ... and when last did you go there?" Linked to territorial instincts, mammalian kinship recognition and the evolved need in all social species to belong, these emotions have deep biological roots. Unconfined to any single region or circuitry in our brains, the emotions associated with displacement and homesickness are rapidly reaffirmed when prodded. Reinforcing these emotion-charged memories is part of the therapy. It is part of the re-experiencing of a primal sense of self and hopefully, a renewed sense of identity.

I believe we all suffer from the same amnesia, the same bouts of homesickness, some of us a little more than others. If there is anyone who vaguely understands the significance of the Aboriginal "walk-about," then you will know what I mean. If you have experienced the chilling night call of the spotted hyena or the symbolic message of solitude in the shape and shade of one of Africa's great trees—the Umbrella thorn, *Acacia tortillas,* then you will know what I am saying or perhaps, more significantly, where I am headed. I think we have all, in our own way, voiced the feeling "I need to get out of here." And I am sure we can all relate to William Butler Yeats' telling lines

> I went out to the hazel wood,
> Because a fire was in my head ...

We have entered what will undoubtedly be known as the Environmental Age. There is hardly a newspaper or news report these days that somewhere in its pages or broadcasts does not draw attention to the growing heaviness of the human footprint on our planet. We are being called upon to re-examine our origins, our status, our language, our economics, our responsibilities and our possible future as a species. We are being challenged, as never before to re-examine our relationship with the earth and with every living creature.

To me, a greater understanding of these relationships is becoming an important part of my work as an analyst. Can personal problems be effectively dealt with outside of an awareness of the impact of the natural environment on the human psyche and vice versa? I don't think so. We shape our environment but we are also shaped by it. It is because of this and because so many of my patients present with identity issues that I now find myself paying an increasing attention to the task of assisting them towards a rediscovery of their biological roots and with it a restructuring of an identity that involves the tempering of the fierce centripetal, self-preserving force of the human ego.

It is important that we understand what the ego is or better still, what it represents. The ego is a psychological "territory." It has "boundaries" without which we become psychotic. Through various denial-oriented mechanisms such as regression, repression, and sublimation, the ego is defended and protected against overwhelming emotions in the same way that we would protect and defend our physical boundaries against uninvited trespassers. From an evolutionary point of view, it is an adaptive, evolving, and necessary aspect of the psyche. Without it, there would be no everyday sense of reality. We would not be able to differentiate between thoughts, emotions, and perceptions. Like a conductor of an orchestra, the ego has an orientating function, coordinating skills such as intellect, perceptions, memory, and emotions. It is a point of reference to what we call a cognitive sense of me versus you. This sense of "me" is what is often referred to as the ego-self and because it is both personal as well as central to one's psychological survival, it is naturally and fiercely centripetal. It is in this light that I doubt that this is the self that Apollo was referring to in his admonition: "Know thyself." Instead, I believe it was an admonition to temper this centralizing force and to seek or create a direction toward a more inclusive sense of self, an Ecological Self— a self that can make sense of and embrace the wild parts of our psyche and of our debt to the natural environment without being overwhelmed by it. To know thyself is to become ecologically literate—to be sensitive to and informed once more, by the wind, by the trees, by the spoor of insects and animals, by the moods of a river, by the potency of solitude ... and by the spoor and dreams of that two million-year-old survivor in all of us. It is to know what it means to be in the skin and the spirit of place. It is a self and an identity that is impossible to define outside of a conscious relationship with the land, with people and animals.

Finally, where in the psyche would we locate this ecological self and how would we know when we had found it? In his masterful essay "The Rediscovery of North America," the ecological writer Barry Lopez writes about "la quarencia," the Spanish bullring chant that describes the place the bull returns to after his painful encounters with the picadors and the banderilleros—that immeasurable yet tangible "spot" in the ring where the wounded animal goes to gather himself before his final charge. I see it as a powerful metaphor for that "place" of the ecological self. It is that *place* where the land and the psyche meet, where one feels real, authentic, elemental. It is that *place,* that spiritual and geographical space where one feels one has come home, where one can gather one's self. It is that inner place, a moveable feast from which one's strength of character is drawn. It is neither a passive nor a prescriptive place and neither do you stumble upon it. You have to find it outwardly and inwardly for yourself . . . consciously. And you must protect it.

And so, as the genetic codes responsible for the variety of the earth's species are slowly unraveled and interpreted, I would hope that our veil of ecological amnesia will also begin to lift. Let us remember where we have come from and that we are the human animal. This is our place. This is our time. We are privileged.

❧

Ian McCallum is a medical doctor and psychiatrist living in Cape Town, South Africa. He is the author of *Ecological Intelligence* and *Thorns to Kilimanjaro.* McCallum has a special interest in evolutionary psychology and the animal-human interface (what we learn about ourselves from animals). He is the Director of Education and Leadership Projects for the Wilderness Foundation Africa and a trustee for The Cape Leopard Trust.

Body as Place:
A Somatic Guide
to Re-indigenization

NALA WALLA

Today, many millions of people are part of a growing worldwide diaspora, a population who may never know precisely where their ancestors lived, or what practices they used to maintain respectful connections to the land. A parallel concept of diaspora—a dispersion from place—applies to those of us who do not feel at home even in our own skins, who feel somehow estranged from our own bodies. Countless modern people are currently wondering: how do we inhabit a true sense of home, an ecologically relevant sense of *place,* instead of a mere *space* where we extract what we need and dump toxic waste when we're done?

Since our bodies are indeed our primary home, any endeavors to create a sense of place must include strategies for getting to know one's body more deeply. Developing and increasing awareness of our own felt-experience can be a beneficial practice that connects us inevitably back to earth via our own flesh. Thus a re-inhabiting of the individual body is an essential step toward *re-indigenization*—behaving in an ecologically respectful and culturally sustainable manner that honors all meanings of "home." The embodied arts are designed to help us to do exactly this.

Zone Zero: Localism Begins with the Body

For those of us in green movements seeking a deeper and more respectful sense of home, it can be helpful to recognize that as infants, our sensation of gravity provided our very first experience of a sense of "place." Our relationship with earth is primary, forming the basis for development of every other movement we make.

150

When we add permaculture terminology to our discussion of "home" and "place," we can refer to the body as the foundation of the concept of *zone zero*—the natural center in a landscape from which all activity radiates. Accordingly, any sound ecological habitat design will plan to optimize bodily health and strive to take basic bodily patterns into careful consideration. Our bodies are the first units of *localism* from which homes, villages, and communities are built.

Ecosomatics: A Working (and Playing) Definition

e • co• so• ma • tics

1. The art of sensing the "inner body" as a way to connect to the greater social and planetary (Gaiac) bodies;

2. The view of somatics as inseparable from ecological health and sustainability;

3. The practice of using somatic principles to facilitate and enhance sustainable work in the landscape, e.g. gardens, farms, village building.

Ecosomatics is an emerging interdisciplinary field that connects movement education, improvisation, healing arts, psychology, ritual, performing arts, and good old-fashioned play with ecological consciousness. The practice of ecosomatics heals the separation between mind, body, and earth by encouraging direct sensory perception of one's body both *in* the natural environment and *as* the natural environment.

In order to understand ecosomatics, we must first grasp the distinction between *soma* and *body:* when you look at me, you can see that I have a body. What you cannot see is my proprioceptive sense (or felt-sense) of my own body—my *soma.* In 1976, Thomas Hanna coined the term *somatics* to distinguish this subtle "inner body" from the outer, gross body. This term is based on the holistic Greek concept of *soma,* which refers to the entire mind-body-spirit system.

Thus, *somatics* as a term encompasses the art and practice of sensing the *soma,* or "the body as experienced from within." The somatic sense is a veritable sixth sense, as it cannot be adequately explained by any of the other five categories: taste, touch, hearing, sight, or smell. In an age where we look to authorities in the human health field other than ourselves, who often dole out conflicting diagnoses and ineffective treatments, *somatics* is an empowering concept that affirms our innate knowledge of our own bodies and encourages us to participate deeply in our healing.

By tuning in to direct sensory experience, we can learn to release habitual tension and pain, while optimizing for ease, efficiency, and enjoyment. Any movement—sacred or mundane—can be re-patterned in this way, from dancing and singing to simply getting in and out of a chair.

We can also apply these sensing and re-patterning skills to activities that aim to restore a sustainable relationship to earth, such as planting trees, harvesting food, and creating a community ritual. Noticing the analogies between patterns in nature and those we sense in our bodies helps us create effective ecological design—as well as better understand our inner workings. When we do this, we venture into the realm of *ecosomatics*.

Re-indigenization: Creating a Sense of Place

> "Okanagans teach that the body is earth itself. Our flesh, blood, and bones are earth-body; in all cycles in which earth moves, so does our body. As Okanagans we say the body is sacred. It is the core of our being, which permits the rest of the self to be. Our word for body literally means "the land-dreaming capacity."
>
> –Jeanette Armstrong, Okanagan Teacher, Activist,
> Traditional Council

The word *ecosomatics* evolved to describe a gateway to the greater "earth-body" via our individual bodies. It expresses the fluid nature of the Self within a world of boundaries at once distinct and permeable. And though the term comes from a scientific language that the modern human—steeped for generations in a rational worldview—can comprehend, the concept is as old as humanity itself, and well understood by traditional peoples.

Because we are microcosms of earthly patterns, practicing respect for our bodies demonstrates respect for the earth, two characteristics of sustainable, indigenous cultures. Members of the diaspora may never be indigenous in the same way as are native peoples who have resided in a particular place or region for hundreds of generations; however, we can (indeed, we must) learn to behave with the same respect for place exemplified by indigenous culture.

A key step toward *re-indigenization* is a "coming home" to our own bodies, a deep "rooting" into earth through our own flesh, a return to the sense of connectedness that is ever-present in sensation. There are many simple somatic practices that can help us access sensation any time we are feeling "out of body." (For starters, see the short resource list at the end of this essay.)

Art as Technology, Not Luxury

At this crucial time in human evolution, it is to our benefit to uproot the foolish notion that The Arts are only for the "talented" and for "professionals," or that they are a luxury the average person cannot afford. The embodied arts are far more than frivolities for the privileged few. These systems were evolved over eons by the grassroots and are best regarded as potent and sustainable "soft" technologies in the truest sense of the word. *(Technology* comes from the Greek word *tekhne,* meaning *skill* or *art.)*

In direct contrast with "hard" technologies (automobiles, computers, etc.), which commonly pollute and consume vast resource supplies, the body-based arts depend only on resources within ourselves. Considering the dangerous imbalance in consumption levels and the over-reliance on hard technology in industrialized societies, a refocusing upon the soft technologies is an intelligent survival strategy.

Technologies of the social realm like dance, ritual, storytelling, and song (the original software!) have always served as communication methods, social exchange and conflict resolution strategies, information transmission schema, knowledge banks, and efficient energy utilization patterns. And, as anyone involved in ecological and activist groups can attest, it is primarily social conflict that impedes our progress and creates burnout, not lack of hard-tech. The Arts serve as grease for the social wheels. Thus, if social dysfunction is the great limiting factor in implementing truly ecological habitats, it is greatly helpful to view them as tools for re-indigenization.

Embodied Activism—Countering Dissociation

As techno-industrial society races along ever more digital and virtual pathways, humans witness the disturbing side-effect of losing touch with our embodied experience. Our modern habit of fouling of our own nest is evidence of a people suffering acutely from disconnection and dissociation from *the body* at every level—the personal body *(Soma),* the social body *(Community),* and the greater earthly body *(Gaia).* Dissociation is a serious psychological pathology yet one so widespread among modern people that it is unfortunately considered normal—and even encouraged by technologies in which we "inhabit" virtual and cyber "worlds."

Since our bodies are quite literally composed *of* and *from* earth (carbon, nitrogen, hydrogen, oxygen ...), a re-inhabiting of our bodies amounts to a profound activist strategy for re-association with Earth and re-indigenization to Place. Aligning with our bodies may seem to be a small contribution, but since our habitual denial of the body lies at the root of our mistreatment of the earth, these small ripples eventually become a sea change that affects the entire world. Simply by becoming advocates for our own flesh and blood, we initiate an *embodiment of activism,* practicing behaviors that come increasingly closer to those of true "indigeneity."

By utilizing the arts, we can facilitate a shift away from slavery paradigms where we push our bodies beyond their capacity, and instead learn to honor our needs for proper alignment, rest, and play as essential to community function. Such a body-based philosophy encourages us to dissolve outdated views of manual labor as a chore that is somehow "beneath" us. Instead, we learn to value earthwork as a privilege: an enjoyable and healing endeavor where we can express our creativity, breathe fresh air, and exercise our bodies—all the while helping the shift toward sustainability.

Beyond the Mat: The Yoga of Earthwork

All the arts originally evolved within the context of Place and Community. Long before they arrived in the halls of Academia or Broadway, the arts belonged to the Folk, who wisely cast them in valuable healing, therapeutic, and integrative roles. For example, somatic practices such as yoga have developed to tune the body-mind system to the daily tasks of building and sustaining a village—squatting and reaching to harvest food, flexing arms and legs to carry water or dig a foundation. To learn how best to push a heavy wheelbarrow and build an earthen house without throwing out our back, becoming bored, or getting sunburned is indeed a yogic practice.

When we recognize the somatic opportunity in earthwork, we move more slowly and deliberately, checking for proper alignment and breath patterns, and we heed our body's requests for a break. By treating our human bodies kindly and humanely, by dancing and telling stories in the garden much like our ancestors have always done, the line is blurred between work and play, between action and activism, between life and art. Over the last few years, I have facilitated many earthwork and building projects that integrate dance, song, bodywork, rest, and play into the worksite and program, with results that are both empowering and fun. (Check out http://bcollective.org/gaia/output5/)

Here I am suggesting that we view embodied arts though a holistic lens. Ours is no armchair movement. We cannot simply sit back while someone else does the "dirty work" for us. To create ecological habitats, we will need to rid ourselves of the outdated stigmas attached to manual labor, and welcome the sweat on our brow. Thriving, sustainable villages and gardens will not build themselves. Only healthy, vital bodies and communities working cooperatively can achieve the vitality we seek.

The arts are designed and destined to move beyond the bamboo-floored yoga studio or velvet-cushioned theater, where they can be put to practical use—out of doors—in our everyday lives. In true egalitarian fashion, the arts have always offered *anyone* (not just professionals) who practices them deep understanding of earth's grand cycles and strength of community. The good news is that these tools are still there. All we need do is use them.

The Ecology of the Body

Recent research in biology has shown that every time we take a breath, a billion electrochemical reactions occur within our bodies, a billion cells are born, and a billion die. This information reveals that the inner ecology of the body mirrors the greater ecology. Our modern sciences have now provided evidence confirming what indigenous cultures have always known: that universal and ecological patterns are right here, beneath our skin, and beneath the soil-skin of the earth. These ecosomatic patterns can be experienced and perceived directly through embodied arts practices.

It is imperative at this time to invest in the soft technologies—the skills, stories, and arts that connect Earth, Body, and Community. As they have always done, the Arts are helping us to create the respectful, cooperative, sustainable cultures of the future, and thus they ought to be widely adopted as best practices, especially among activists. And as a pleasant side effect, our lives will include more play and more celebration as we become, once again, indigenous to the places we live. As Dr. King once said: even if I knew that the world would go to pieces tomorrow, I would still plant my apple tree today.

LINKS AND RESOURCES

Hanna, Thomas. *Somatics: Reawakening the Mind's Control of Movement, Flexibility and Health.* 1995. New York: Dacapo Press.

Hartley, Linda. *Wisdom of the Body Moving: An Introduction to Body-Mind Centering.* 1995. Berkeley, CA: North Atlantic Books.

Zaporah, Ruth. *Action Theater: The Improvisation of Presence.* 1995. Berkeley, CA: North Atlantic Books.

The Feldenkrais Method of Somatic Education Web site www.feldenkrais.com/method/the_feldenkrais_method_of_somatic_education/

The Open ATM Project (Awareness Through Movement), free Feldenkrais exercises for download: www-ccs.ucsd.edu/~falk/openatm/

Feldenkrais Exercises from the University of Utah Somatics and Human Development Laboratory: www.psych.utah.edu/lab/somatics/ex-try.php

Daily Somatic Essentials—"the catstretch" from Clinical Somatics Web site: http://somatics.org/shop/guides/catstretch.html

The School for Body-Mind Centering: www.bodymindcentering.com/About/

❧

Nala Walla is a transdisciplinary artist, educator, and homesteader in Washington State. Nala holds a master's degree in Integrative Arts and Ecology, and is a founding member and facilitator of the BCollective: an umbrella organization dedicated to creation of healthy and sustainable culture through the embodied arts. The Bcollective offers community-building workshops, creates participatory educational performances for kids and adults, and hosts permaculture skill shares from ecosomatic building to creative mediation. Please visit www.bcollective.org for more information, or contact Nala directly: nala@bcollective.org.

Morality Is
a Somatic Experience

TOM MYERS

When I was about seven, I captured a frog from a little pond in the woods behind David Rice's house. I carried it home in my pocket, a dry and cramped space for a small amphibian, then I proceeded to torture it in other ways, culminating in hitting it on the head with the mace of a spiky horse chestnut casing. I know where the impetus to do this came from—in rural Maine in the 1950s, the older boys in the neighborhood were always chasing or shooting some animal, and I wanted to be one of the gang. The feelings of animals—feelings at all—were not high on anyone's list. This time, though, I was alone: bringing the frog home and hurting it was my own project.

But where did the other voice come from? The voice that suddenly saw this for what it was: a pointless exercise of power, the infliction of useless pain, the epitome of unfair advantage. The voice could have issued from my parents, or from God, but I actually felt it from inside my own body. Somewhere in my chest, between my lungs and behind my heart, a light came on, and the thing that I call "me"—the perceiver—relocated.

Heavy with remorse, not even recognizing the boy who had blithely separated himself from feeling, I carefully carried that particular little green frog—one of hundreds in that pond, one of thousands of ponds in Maine—back into the pine wood to set it gently in the water, and I waited while it regained its equilibrium and suddenly shot away from me below the surface.

What kind of change are these "visceral" experiences, these times of an internal "wrench" that relocates the spiritual self within the body? This question turns on our view of what a body is.

Twenty-First-Century Monism: Bodymind

In Western culture, the body has been equated philosophically with our lower selves, especially since Descartes' deal with the church that left the body in the realm of object (a "soft machine") and the mind in the realm of the immortal soul, God, the cool rarified atmosphere of heaven. The body is animal: dirty, mortal, tricky, full of uncontrollable urges, messy in its needs for food and evacuation, venal in its need for sex, and, in a popular interpretation of Genesis, the source of sin.

In the great Chain of Being, we are poised above the animals but below the angels. The body is the lower, animal part, and the mind is the noble, celestial part. Virtue is defined as our ability to control and dominate these bodily urges (and by extension control and dominate nature to "civilize" it). We laugh now (or cry) at the nineteenth-century presumption of bringing "civilization" by clothing and Christianizing the "savages" of the Americas, Australia, and Africa, but that attitude remains even now between us and our bodies. This dualism is an extraordinarily strong cultural underpinning to our perception, thinking, feeling, and action.

In fact, your body is a source of wisdom and counsel, as well as a tool for deep spiritual communion. All spiritual practices are essentially tools for producing particular bodily states. This has always been around, in the soft darkness just outside the campfires of the Children of Abraham, in the form of animism, Taoism, and Tantra—philosophies that value the corporeal self in its wonder and magic and innate knowledge. Now we are urgently called upon to apply such philosophies in a modern context, not to deny the intellect but to marry it to its base. It is a hard concept for the modern to get, but I contend that there is no mind; "minding" is a function that bodies perform. We need a new monism that marries body and spirit, to counter the dualism that has brought us to this pretty pass.

The challenge of the twenty-first century—and it applies to problems of war, the environment, and cultural survival—is: how do we educate a Neolithic body to live successfully in an electronic world? The genetic basis of the body has changed little since we first domesticated fire and painted on cave walls, but during that time we have altered the environment we live in to an almost unrecognizable degree.

Every generation thinks that its own crises are greater than those faced by their forebears. In the clouds looming over inhabitants of this century, however, there really are telling differences. With the fully global nature of humanity's interactions these days, there is no frontier left, no room for dumping the results and moving on to the next green pasture. You cannot throw something away—there is no "away." We have shown little propensity for tackling our problems with courage, energy, imagination, and initiative. Bucky's "spaceship earth" hurtles on through space, most of its inhabitants still unaware that they are part of the crew. The consumptive attitude that permeates the West and its economies is clearly unsustainable yet shows little sign of abating. The alienation from the body and the "natural" is reaching epidemic proportions. There is cause for pessimism.

Yet we tend to believe that technical solutions exist for all the problems confronting us—energy, pollution, quality of life, read Amory Lovins—and the young are full of boundless enthusiasm and conviction that all such challenges will fall before our ability to organize solutions. There is cause for optimism.

Physical Education in the Twenty-First Century

On a practical level, a new comprehensive physical education is key to the challenges that face us. As a somatic therapist and teacher, I believe that over the medium term each of these puzzling locks (energy, politics, the environment, and personal plasticity) can be picked by an increased and focused awareness on the physical self and its complex relations with the world, others, and its own essence—the kind of deep inner experience suggested by my opening story about the frog. How can we cultivate the deep self-sensing that leads to authenticity in ourselves and in the young?

Although some of our kids are challenging themselves on the planet as never before—on sea, snow, and air—the general running down of physical capacity is frightening. Everywhere in schools, physical education is being dropped, cut, ignored, and abandoned. What educational programs remain are firmly based in an industrial approach to physical culture, centered on repetition and competition.

Repetition is fine for habit-setting purposes, and competition is fine for building performance and good sportsmanship, but are these the basis for a twenty-first-century physical education? Repetitive moves in a competitive atmosphere are tailored to prepare a generation for jobs in an industrial world, often interacting with a machine in the service of production. In fact, each generation fashions its physical education to the world it finds (see the brilliant film *Dance and Human History* at www.css.washington.edu/emc/title/719). For example, cricket was perfectly suited to the running of a vast, slow empire in the last century, baseball to the entrepreneurial culture of the 50s, football to corporate culture of the 80s, and soccer to the cheerful anarchy of European socialism.

In the twenty-first century, however, anything repetitive—from jumping jacks to writing romance novels—can be done better by machine. So we need to teach our children to explore what is original and individual in their movement, not restrict it to mindless repetition.

For a new physical education, we must cast a wider net than do the current myths of "exercise." It is great when the inactive become more active, and more time devoted to even basic physical activity in schools would be welcome: *mens sana in corpore sano,* as Juvenal said. Exercise, however, is but one thin slice of what a new physical education could provide.

In the first, pre-verbal year, a tremendous amount is conveyed to children in how they are physically handled. You cannot talk a baby in and out of diapers, clothes, or car seats. The non-verbal dialogue of guided movement in the first year underlies the dialogue you have with him or her as a teenager. Courses in baby handling offered to all parents could transform parenthood into a richer syntax between child and parent (especially fathers, who in our culture often know next to nothing about handling and communicating with tiny bodies).

Understanding the natural spiral movements that lead into easy, upright alignment would eliminate many of the aches, pains, and degenerative diseases that plague and cost our society. Understanding the somatic aspect of feelings transforms the perils of adolescence. Stress, however much you may touch it with talk therapy, is a fundamentally physical, bodily response. The ability to detect stress within the body facilitates less conflict in social situations, as well as reducing chronic disease. Maintaining movement into senescence can extend productive lives. Most kids graduate from high school knowing more about the principal exports of Chile than they know about their own feelings and their bodily language.

A well-constructed program could build intuition, which is grounded in our bodily "hunches" and kinesthetic perceptions. In fact, there is a vast, largely unstudied area we could call "KQ," kinesthetic intelligence or physical intelligence. We are familiar with IQ (Intelligence Quotient) and becoming familiar with EQ (emotional intelligence), but KQ is largely uncharted. In my own case, I have two left feet for dance but have significant KQ in my hands—a fact I discovered when I was eight but did not recover and utilize until my late twenties, when I encountered meditation and the martial arts, and took up manipulation of bodies as a career. Other people have startlingly good contact with their intuition but no ability to relate socially. The kids doing lay-ups in vest-pocket parks in New York display highly developed KQ, which is largely unused except by those who end up with NBA contracts.

Living now requires us to return to the lived experience of the body. I believe that doing so would produce a human who is more aware and more compassionate (as in the story with which we started). The physically educated human can be more calmly energetic, more sustainable, more aware of what is eaten and what is pooped. The basis of ecology is in the body; the basis of peace is in the biological cooperation.

Would a kinesthetically evolved populace tolerate injustice? Send its children off to war? Madly and blindly consume? Tolerate so much waste?

Our commercial culture pays billions (in advertising) to keep us away from our true sense of ourselves. To be constant consumers, we must constantly feel a lack: you don't smell right; your hair is less than perfect; you need this medicine; if only you had this car, you would get a makeover and a smile. . . . It is a distasteful and unsustainable way to build an economy, and it is very bad for our bodies and biosphere.

However we get to a sustainable system, it is the only choice available in the long term. Continue on our merry way, and we will be eliminated, another experiment tried and failed by the sun playing upon the earth's rich surface. We are not essential to the process; we are just the current best hope. The whales and elephants may be intelligent, but they do not hold the future of the planet in their flippers or trunks. We do—sad, but true. Grounding ourselves back into our bodies will help keep us sober but joyful in the quest to hatch into the light of day from the planetary eggshell that has surrounded us from our inception. Then we can be conscious stewards of this, the most beautiful planet we know.

More specific ideas along this line can be found in the "Spatial Medicine" section of the www.anatomytrains.com Web site.

If we are indeed looking down the barrel of environmental catastrophe, as the question that launched this book project assumes, the depth of that barrel is the degree of our alienation from our physical selves.

How to Live Now

Here's a little advice from the point of view of thirty-five years in the somatic therapy craft:

- Get out and move. All forms of exercise are helpful, though walking, running, swimming, and singing carry the longest evolutionary history in terms of healing. "Walking your blues away" is more than a line in a song. Four million years of upright walking have ensured that this exercise has many salutary benefits to the body and its minding.

- Take on a bodily practice—Tai Chi, yoga, Pilates, chanting, meditation, martial arts, rock climbing, and any of a hundred more now on offer in your community.

- Reconnect with your gut feelings, your hunches, your subtle feelings, your physical trust, your hungers, your blood dreams, your bitter tears, your gasping sobs, your laughter, your heart's desire, the smell of possibility, the taste of success, zero at the bone, even the voice of reason. That which is most reasonable in us is the part that does not reason—the part that is growing your hair, repairing your liver, and digesting your food. Teach your children to trust their gut.

- Use your movement to explore your genetic history, your future possibility, your capacity, your alignment, your balance, your power, or just the sheer joy of the ability to shapeshift.

- Somatic therapy (bodywork) involves someone else's eyes and senses on your body, to help you find "lost" parts of yourself, areas subject to "sensori-motor amnesia" that only an outside pair of hands can help you discover, because they have fallen out of your body image. Alexander Technique, Feldenkrais, Structural Integration, and any of a hundred more—the practitioner matters more than the method—can help you recover your complete self. The best decisions are made when all of you participates.

- Make sure your children are getting more and more in touch with their bodies, learning to rely on their authentic feelings. Move with them: roughhouse, wrestle, hug, rest next to each other. Monitor the physical education that they are getting in their school. Keep the bodies of your children growing in their KQ, their capacity to feel and respond to the challenges around them. Help them not to succumb to sedentary slavery to the electronic world or the numbing alienation of false gods, consumer goods, and outer goals that will do little to make them truly happy.

And then, fully mindful (which we now know is "bodiful"), get up and get to work! There is so much to be done at every level, at every turn, on every face.

Tom Myers is the author of *Anatomy Trains* (Elsevier, 2001, 2009) and numerous trade and journal articles. A student of Ida Rolf, Moshe Feldenkrais, and Buckminster Fuller, he has practiced integrative bodywork for thirty-five years in the U.S. and Europe. Tom lives with his partner Quan on the coast of Maine where he directs Kinesis, which offers training in manual therapy and the anatomy of movement worldwide. His Web site is www.anatomytrains.com.

Earth Rights

DR. VANDANA SHIVA

The collapse of Wall Street in September 2008 and the continuing financial crisis signal the end of the paradigm that put fictitious finance above real wealth (which is created by nature and humans), profits above people, and corporations above citizens. This paradigm can only be kept afloat with limitless bailouts that direct public wealth to private rescue instead of using it to rejuvenate nature and a public economic livelihood. It can only be kept afloat with increasing violence to the earth and people. It can only be kept afloat as an economic dictatorship.

This is clear in India's heartland, where the steel and aluminum corporations' limitless appetite for profits is clashing head on with the rights of the tribals to their land and homes, their forests and rivers, their cultures and ways of life. The tribals are saying a loud and clear "no" to their forced uprooting, and unrest is growing. The only way to get to the minerals and coal that feed the "limitless growth" model in the face of democratic resistance is the use of militarized violence. Operation "Green Hunt" has been launched in the tribal areas of India with precisely this purpose, even though the proclaimed objective is to clear out the "Maoists." Under operation Green Hunt, more than forty thousand armed paramilitary forces have been placed in the tribal areas. Operation Green Hunt shows clearly that the current economic paradigm can only unfold through increased militarization and the undermining of democratic and human rights. The economic fundamentalism of "limitless growth" is clearly collapsing. The technological fundamentalism that has externalized costs—both ecological and social—and blinded us to ecological destruction has also reached a dead end. Climate chaos, the externality of technologies based on the use of fossil fuels, is a wakeup call that we cannot continue on the same path. The high cost of industrial farming is running up against limits, both in terms of the ecological destruction of the natural capital of soil, water, biodiversity, and air and in terms of the creation of malnutrition, with a billion people denied food and another two billion denied health because of obesity, diabetes, and other food-related diseases.

We need a new paradigm for living on the earth, because the old one is clearly not working. Finding alternatives is now imperative for the survival of the human species—not only alternative tools but alternative worldviews. How do we look at ourselves in this world? What are humans for? Are we merely money-making and resource-guzzling machines? Or do we have a higher purpose, a higher end?

I believe we do.

I believe that we are members of the Earth family—of Vasudhaiva Kutumbkam. And as members of the Earth family, our first and highest duty is to take care of Mother Earth—Prithvi, Gaia, Pachamana. And the better we take care of her, the more food, water, health, and wealth we have. "Earth rights" refers first and foremost to the rights of Mother Earth and our corresponding duties and responsibilities to defend those rights. Earth rights are also the rights of humans as they flow from the rights of Mother Earth—the right to food and water, the right to health and a safe environment, the right to the commons, the rivers, the seeds, the biodiversity, the atmosphere.

I have given the name "Earth Democracy" to this new paradigm of living as an Earth community, respecting the rights of Mother Earth.

Earth Democracy enables us to envision and create living democracies. Living democracy enables democratic participation in all matters of life and death—the food we eat or do not have access to; the water we drink or are denied due to privatization or pollution; the air we breathe or are poisoned by. Living democracies are based on the intrinsic worth of all species, all peoples, all cultures; a just and equal sharing of this earth's vital resources; and sharing the decisions about the use of the earth's resources.

Earth Democracy protects the ecological processes that maintain life and the fundamental human rights that are the basis of the right to life, including the right to water, the right to food, the right to health, the right to education, and the right to jobs and livelihoods. Earth Democracy is based on the recognition of and respect for the life of all species and all people.

Ahimsa, or nonviolence, is the basis of many faiths that have emerged on Indian soil. Translated into economics, nonviolence implies that our systems of production, trade, and consumption should not use up the ecological space of other species and other people. Violence is the result when our dominant economic structures usurp and enclose the ecological space of other species or other people.

According to an ancient Indian text, the Isho Upanishad:

> *The universe is the creation of the Supreme Power meant for the benefits of [all] creation. Each individual life form must, therefore, learn to enjoy its benefits by forming a part of the system in close relation with other species. Let not any one species encroach upon other rights.*

Whenever we engage in consumption or production patterns that take more than we need, we are engaging in violence. Non-sustainable consumption and non-sustainable production constitute a violent economic order. In the Isho Upanishad it is also said:

> *A selfish man over-utilizing the resources of nature to satisfy his own ever-increasing needs is nothing but a thief, because using resources beyond one's needs would result in the utilization of resources over which others have a right.*

On Earth Day 2010, the President of Bolivia, Juan Evo Morales Ayma, organized a conference on the rights of Mother Earth. The intent of the conference was to start a process for adopting a Universal Declaration of the Rights of Mother Earth, similar to the Universal Declaration of Human Rights. The idea came out of the failure of the Climate Conference in Copenhagen, where Morales had said, "If the earth is recognized as Mother Earth, it's something that can't be sold. It is something that can't be violated. It is something that is sacred. This is ... why I've come here: to defend the rights of Mother Earth, to defend the right to life, and to defend humanity."

Without Earth rights, there can be no human rights; Earth rights are the basis of equity, justice, and sustainability.

Earth rights are human rights.

⁊

Dr. Vandana Shiva is a world-renowned scientist and environmental activist, well-known for her work in the fields of biodiversity, genetic engineering, and water rights. A member of the World Future Council, she is also the author of several books, including *Biopiracy, Stolen Harvest, Water Wars, Soil Not Oil,* and *Earth Democracy,* from which portions of this essay were adapted. Her Web site is www.vandanashiva.org.

Indigenous Mind

KAYLYNN SULLIVAN TWOTREES

A s I sit overlooking the valley where I live, it takes a moment for me
to still myself enough to allow my breath to settle deep in my belly. My
gaze drifts over the valley and I can see the road I traveled to reach this
viewpoint.

Details of my day, my life, my worries, my longings are like the single fruit trees I see from my perch on this ridge. The longer I sit the more
the edges of my thoughts and the specifics of the landscape blur until I
can feel myself cradled in the lap of the Topa Topa Mountains that caress
the edge of the valley. In the midst of those blurred edges I can hear the
ravens and hawks more clearly above me and the bees in the rosemary
bushes. I can feel the power of the mountains and the almost imper-
ceptible movement of the earth below the mountains as well as the sweat
on my upper lip heated by sunlight. The wind comes to me through
the movement of the trees and my body responding with my breath.

I am a detail of this small valley and I am linked to the earth through
the movement under the mountains. I can sense a deeper presence than
I can imagine, softening the boundary between known and unknown
that is made hard-edged through reason. Each creature in my locale
speaks to me—bees, birds, ants, dogs, trees, flowers, humans and their
machines, as well as the beings I can't identify. Through these messages
I can sense the web of connection through my body and my breath. I
awaken to the subtle layers of messages from this specific place to the
planet and into the unknown.

No matter how much I long for an imagined future for myself and
the planet, or drag myself back to a perceived idyllic past, I am still breathing into the "right now." Longing, worry, agitation, and frustration can't

change the fact that this moment is the one before me. It is the one I can experience and transform. This moment I can choose to be right where I am in the fullest possible sense. I can choose to be aware of the wonder, grace, problems, emotion, connections, conflicts, joys, and pain of it. And I can breathe to open my senses so that I can feel myself as a small part of the moment in preparation for the depth of possibility it holds. I can choose to feel my breath linked to a breathing planet through trees and plants. I can expand my listening to hear the sounds of the other beings who are sharing the moment with me so that my mind does not fill all the space.

This awareness is not a state of achievement. It is like walking. I take a step and as one leg lifts off the ground I am close to falling. As I set my foot down again in the movement forward I find stability. My stability is measured by my orientation. With each breath I am on my way to falling into the whirlwind of human-created stimuli- messages, information, emotions, obligations. I can miss out on the moment with an amnesia that highlights my emotional response and obscures the wonder of the moment. I can also choose to breathe and create space to look beyond myself to my North Star and reorient myself to my aspiration—relationship and connection. Or I can choose to tentatively open my senses and my body. I can expand my moment to the landscape of which I am a detail and find inspiration in the trees, plants, minerals, bodies of water, and creatures of land, air, and water around me. I can be reminded by nature and become aware of my breath in the moment with my *indigenous mind.*

Our reality is held in place by agreements with time, space, and mind. Magic, medicine, and the creative impulse are the means by which we can renegotiate one or more of these agreements.

Through the automatic function of breathing, the work of the hands, words sent on the breath as sacred language, and the spaciousness of silence to engage the unknown we can transform both understanding of self, *I-dentity,* and the shifting web of relationship, *We-dentity.*

The breath is a coalescence point, geographic, metaphysical, and physiological with access to multiple dimensions of possibility. In the midst of inhale and exhale there is a point in space, time, and mind where we can meet the tension of intersecting perceptions with curiosity, wonder, and the willingness to let possibility overflow into the moment.

What generates hope and gratitude is the experience of each moment with a willingness to find possibility more invigorating than fear of the unknown. At each crucial point of intersection we can listen for the subtle ley lines where miracles are nourished.

The tension of intersecting perceptions is the generative energy that breaks open seed coats of consciousness and makes visible simultaneous multiple realities. Living right now with the earth as she changes with me and around me, intellectual-emotional turmoil and currents of possibility intersect. This intersection creates the tension to break open my *indigenous mind.*

What I am invoking by *indigenous mind* is the power of re-energizing our world with all of the hundreds of senses that open our awareness to the web of relationships that are the earth. The power of this consciousness to renew, adapt, and regenerate in new forms is without question more powerful than our single species. Accessing this means a commitment to slowing down, remembering, and re-conceiving on a smaller scale to reclaim intimacy with nature and its layers of species as an aspect of my own essential nature. In this way I remain aware of being uniquely indigenous to this planet through space, time, and mind. *Indigenous mind* is the innate ability to become aware of the earth at an intimate and dynamic level and respond to the messages and stimuli from the beings and life in the moment.

All humans are indigenous to planet Earth and every one of us has the ability to awaken *indigenous mind.* This is not a cultural lens but an individual awareness of our moment-to-moment presence as details of our locale. We can open ourselves to awareness of the intricate links in our species, with a collection of lesser differences (race, culture, gender, geography), hosted, nourished, and impacted by our relationship with the planet and all other species.

This does not mean that our perception of these lesser differences may not color or affect power dynamics and the ideology of oppression.

It does offer the possibility for greater discernment about those perceptions and broader accountability for uses of power. It offers species relationship and interdependence as a means to securing a future place on this planet along with the other life that sustains us.

Traditionally the cosmologies and social structures of indigenous cultures held the earth as sacred and developed the capacity to respond to multiple stimuli from both the landscape and other species. The wind, water, fire, rock, earth, four-leggeds, birds, insects, plants all sent messages. These cultures taught and modeled that humans were able to respond to these messages as details of the landscape. The teachings of indigenous peoples abound with these practical applications of relational thinking and subtle distinctions of interdependence and difference.

These indigenous wisdom traditions were born out of deep understanding and connection to cultural or environmental relationships to specific geographic locations. In many indigenous languages the names of different nations, tribes, and sub-tribes held both the idea of humans and their relationship to a specific place or geography (the river people, the mountain dwellers, the people who scatter their own). These languages held the names of the people beside the names of the other nations or species (the stone people, the standing people) defined and framed by the immediate geography. Difference had everything to do with relationship to the specific place.

In an evolving world, living now calls us to evolve our understanding of "indigenous" beyond culture and specific locale to an awareness of it as the basic link of consciousness in all life including humans and our planet Earth. *Indigenous mind* is an evolution of awareness of our natural state of relatedness. Opening to it shifts our perceptions of difference. It calls us to adapt our orientation and build our individual and collective capacity for thinking relationally. *Indigenous mind* enhances our ability to move from the micro: our particular locale to the macro: the relationship of facet or specific to the Whole.

What we need is a new orientation for the cartography of interdependence. We need compass points to align us with the larger web of relationship, our specific place in the moment and the mechanism that corrects our course. *Spiration* is our orientation and is linked to the automatic function of breathing. Aspiration is the North Star that lifts our gaze upward and guides us to remember the larger web of relationship. Inspiration is our inward awareness of that guidance in our moments of choice available at each breath. Respiration grounds us in the details of our lives. AIR.

My nature binds me to the earth as tightly as I am bound to tree and plant producing oxygen, not only for physical survival, but for the awareness of myself as part of the planet and her changes—not as cause or solution, but as an intimate and dynamic detail of her process. I can fall prey to the rhetoric and drama of issues, events, and emotions or I can breathe into the tiny moments of choice that allow me to collaborate and participate in my own unique way as a partner with the earth throughout the web of relationships with all other species.

I can continue to ask "how do I live my life right now?" or I can take a breath and remember—to listen as well as talk to nature, to receive the breath of the earth through plants and trees, to feel the smallest vibration of relationship in our actions—to redeem my *indigenous mind* and move from human centered thinking to earth centered perception. We have the power to live our lives right now with a new orientation—to listen to the layers of species on the planet for their deeper messages; to be inspired by familiarity and intimacy with our neighboring species; and to open our powers of perception for the possibility for revealing a common future.

❧

Kaylynn Sullivan TwoTrees has spent a life at the crossroads where species, cultures, beliefs, and the unknown collide and find both dissonance and resonance. Based on the Seven Directions Practice that she developed over the course of twenty-five years, and with the input and guidance of indigenous elders, her current work, Practice for Living/Living Practice, helps humans re-orient to our indigenous mind and regenerate our essential relationship with the earth's wisdom. For more information about her work visit www.ktwotrees.com.

HOPE BENEATH OUR FEET

CHAPTER SIX

Balanced Engagement

Wonder:
A Practice for Everyday Life

MUNJU RAVINDRA

When driven to the brink of despair by heartbreak—whether personal or planetary (the despair, in my experience, *feels* the same), I take it as a kind of daily practice to notice, to bear witness, to look. Look. Again.

The way midday light shines through a drooping fern, so that the spores underneath transform into stars. The dark glossy "waiting-ness" of an empty pre-dawn parking lot. Rain drops clinging to the undersides of branches, lined up like pearls waiting to be plucked ... They will dry when the sun emerges, but in the meantime, I uncover their ephemeral beauty, simply by noticing it. By looking. By looking again.

I marvel also at the whimsical efforts other humans have made so that I, so that *you,* can experience delight: the meticulous placement of stones across a neighborhood stream; the quirky grottos and rock seats in an urban greenhouse; the raucous display in a downtown toy store window; the careful preparation of an elegant meal. We build these gardens because we know, by some kind of prior knowledge, that even in a world made completely virtual, the key to our survival is wonder. I think often of Thoreau's famous line, now gracing t-shirts sold in earnest coffee shops: 'in wildness is the preservation of the world.' Let us dare, for a moment, to rewrite the master: *in wonder is the preservation of the soul.*

That's a mighty task for a little word, and frankly, a mighty task for you and me. Caught up in our work, in our activism, in our endless efforts to put the world right; caught up, even, in remembering our own selves; it is easy to walk past the puffed-up spring chickadee balancing awkwardly on the end of a branch. But when we look at him anew, with attention, we can see the world for a moment through the eyes of innocence, and magic starts to blossom all around us. It is this, I believe, that is the art of worldly wonder—it is an *attitude* to daily life. I think of it as a kind of *yoga* of everyday life.

Why Wonder?

Sometimes, for an instant or two, I worry that wonder is a copout. In the face of environmental catastrophe, climate change, torture, famine, and war, what does wonder do to change the world? It is not, at first blush, a glaringly activist choice. But, working on wonder has several important effects:

For one, it re-instills in each of us a sense of what is "true," thereby enhancing our resilience in times of crisis, sorrow, and the general impossible-ness of getting things done. For me, wonder is a flush feeling—a sensation of enlarging, of filling with space, of making room for experience or revelation. I suspect this is how some folks encounter God. Wonder connects me to something larger than myself and gives me the energy I need to keep on agitating. It also gives me the reason.

Working on wonder stretches and strengthens our "wonder muscle," enabling us to pass along the gift. As a child, I was taken to Europe, to India; I saw poverty and despair as well as the extraordinary manifestations of human creativity. At home in Canada, my mother encouraged me to run un-fettered through the woods behind our house, didn't blink when I spent whole days lying on my stomach examining the intricacies of lichen, and loaned me her canoe so my best friend and I could explore the coast. This past Christmas, she presented me a copy of Rachel Carson's *A Sense of Wonder,* with the following note: "with love and gratitude for how *you* have helped *me* to see the world of nature." There's a lesson for all of us in that note: sow the seed. When your muscles tire, some other practitioner of wonder will have grown up to hold your hand. It is, I believe, the greatest gift a parent can give a child.

A life with wonder is simply more fun than one without. Remember those days when you woke feeling that nothing could go right, and then something comical or beautiful or strange caught your eye and you looked, and looked again? It only takes an instant, but it is transformative.

A Daily Practice

We all experience moments of grace, transcendence, awareness. Our challenge, however, if we want to keep "living now" when we think the world is going to hell in a hand basket and we're certain all hope is gone, is to find ways to *create* those moments. Everyday. This doesn't sound like an easy proposition. How do I even *remember* to wonder, I wonder? But it's not that hard. It requires of us the eyes of a child, of a beginner, of a newbie. And we've been those things before so we have a great deal of experience to draw on. It requires of us simply to look. To look again.

Practice, unsurprisingly, is the key. Here's a personal confession: I live in wonder. Not always. Certainly not twenty-four hours a day. But daily. Sometimes I have thought maybe I was born that way, that perhaps it came from my upbringing, or the jobs I've had, or is a strange accident of chemistry and circumstance. But I now realize that it is because I make wonder a daily practice. Especially when I'm feeling down or blue or lost, I just buckle down and practice wonder.

My Wonder "Workout"

I spent many years working in the "sense of wonder business," as an interpretive naturalist in Canada's spectacular national parks. My job was to inspire in park visitors a sense of wonder about the natural world they were visiting, the idea being that one protects what one loves and that, with their latent sense of wonder awakened, these same visitors would form a constituency of conservation for the national parks themselves, as well as for nature as a whole. It's a fine idea, and one that, sadly, is rather difficult to justify in meetings about budgets and national objectives.

Nevertheless, through that work I discovered some basic tools that may help you as you try to "live now," with wonder and awe in your daily life. I offer them to you here as a kind of practical workout for wonder. Some of them are a bit goofy, but give them a try. If you have other wonder-building exercises, tell me.

Investigate the world upside down. You can do this by hooking your legs over a railing and leaning back, dangling from a tree branch if you're young enough or strong enough, by standing on your head like a yogi, or any other way you can think of. Do it for long enough that this vantage point becomes your new "normal" perspective—images compressed into a thin band at the top of the picture (trees, laundry hanging on a line, cars going past).

Surprise yourself. Leave yourself a sticky note. Seed your living space (and your coat pockets) with oddities, curiosities, things you might forget about and then encounter. Choose something that won't mold, melt, or rot. Chocolate is enticing, but impractical. You need something like an unusually colored pebble, an acorn, a piece of beach glass, a bottle cap, a matchbook from a nefarious place you visited, the seed of a particularly curious plant.

Spend the day with a mystic, lunatic, or writer. Or, for that matter, a child (who, if schooling and society don't manage to weld shut his door to amazement, will no doubt one day become a mystic, lunatic, or writer). These people have their heads screwed on sideways and hobble around gobsmacked by the beauty and despair of the world. If you opt to spend the day with a child, try to find a small one, preferably raised by hippies on a commune on the coast, but really any child will work, if you actually pay attention to what *they* have to show *you*.

Lie on your back somewhere other than your bed. Outside, under a tree would be great, but any place will do. The middle of your kitchen, for example, or a busy sidewalk in your town's urban core. Stay a while. Look. Notice, if you're under a tree, how it is like a giant, branches outstretched like arms, holding up the sky. What else do you see?

Let it happen. Sometimes, the most awe-inspiring moments are those that creep up when we're quiet and listening. A fellow naturalist and I used to take turns leading night walks in a national park in eastern Canada. We would be out for hours, without flashlights, guiding our visitors along

HOPE BENEATH OUR FEET

the trail, through the forest, through their fears, feeling our way by the sound of our feet on the paths. At one point, we'd sit down along the trail, close our eyes, and just listen, quiet the body, and attempt, albeit falteringly, to quiet the mind. We invariably had extraordinary experiences—a coyote howling a hundred feet away; a family of mice scampering through our midst, so certain of our *belonging* in that forest with them that for once, faced with twenty humans, they felt no fear. My colleague returned from his walk once flushed with excitement: a majestic moose had walked right through the middle of his group as they had been sitting, *open to experience.*

Get uncomfortable. Even those of us in the "biz" sometimes lose our senses of wonder. Months of answering persistent "what is it?" questions can erode even the most genuine wonderer. So, when the world presses in with its schedules and plans and anxieties, I do something a bit yucky—let a slug from my garden crawl up my arm, or lie down outside in a terrible rainstorm. It always works. Once I get over the instant "you're going to get dirty/catch a flu/be slime" reaction, I begin to feel amazed. When you've been pummeled by rain for half an hour, or have slug trails drying sticky on your skin it is hard not to feel at least a kernel of awe.

Stay up all night, outside. Maybe it's a meteor shower in August, or spotlights glancing off a skyscraper, or the sound of spring peepers in a nearby pond. Whatever you find in the places near you, night will give you an entirely different view of your world. Remember that a missed night of sleep is only that, whereas a moment of true wonder will last the rest of your life.

Go Alone. Whenever you can, go alone. Find a way to get alone for ten minutes every day. Walk. Wander. If you can clear your mind and let the outside in, wonder will find you.

The easiest route into wonder is to shift your perspective. We have all experienced those blinding moments of clarity when adrift in a new city, in an airplane looking at the patterns below, or on a boat gazing back at the land. And we can experience those moments here, now, in our daily lives, by cultivating a habit of attention. It is the lesson of the ecologist, the photographer, the poet: Look. Look again. Wonder.

Munju Ravindra is a naturalist, storyteller, gardener, planner, ecologist, designer, explorer, wonderer, and observer of beauty. After many years of working in national parks, she now works as a consultant who helps parks, museums, and businesses create opportunities for "transformative experience." More information about Munju and her work can be found at www.ideasunlimited.ca.

Embodying Change

CHERYL PALLANT

Between 1998 and 2003, I made several trips to Malaysia to teach. When not teaching, my sights were set on taking a train to the Taman Nagara National Park. Covering more than 1,600 square miles in Peninsular Malaysia, this area is home to 10,000 plant species, at least 250 bird species, wild ox, tapir, elephant, leopards, tigers, numerous monkey species, and the hairy rhino. The park provides guest hideaways that are semi-camouflaged wood perches in the trees for observing the life abounding in the underbrush and among the branches.

The hairy rhino caught my attention because this shy, solitary, plant-eating mammal teeters on the brink of extinction. This rhino, the smallest of its species, grows no taller than four feet and no longer than ten feet. Under cover of night it moves quietly through the thick jungle foliage, feeding on sweet saplings, mangoes, and figs. Threatened by deforestation and poaching, the rhino population here has shrunk drastically in a ten-year period, from thousands down to a few hundred. Recently I read that ecologists trekking into the national park found no hairy rhinos. These docile creatures, the author reported, were extinct. My eyes welled with tears from overwhelming helplessness, anger, and despair.

Unfortunately, in the last several years, such feelings have become commonplace. One need not go halfway around the world to encounter dismal realities. On any given day, a conversation or news report about climate change, extinction, human rights abuses, violence, political chicanery, disease, racism, poverty, arms sales—an endless list of human-generated atrocities—reveals that suffering takes place in all countries. Any one incident can arouse gloom and defeat in me when the world's weight seems an insurmountable burden. What can I do amid so much suffering? Can any of my actions make a difference?

As a poet, often my response to suffering is to write. Reflections on genocide birthed a series of poems, *Into Stillness.* The unspoken emotions and hidden traumas of survivors found voice on my pages. The book is for them; it is also for me. Writing renews my spirit. Cultivating image and phrase and rhythm as they arise from my being affirms my participation in a generative life. Creating is an empowering act. It lays bare the ongoing, awe-inspiring cycle of life and death. Yet the gratification from my writing didn't last.

Other actions resulted: writing letters to Congress, signing petitions, participating in demonstrations and vigils, donating money, volunteering at a phone bank, performing rituals. No amount of activity loosened the stranglehold. What more could I do without burning out and negatively affecting my job, my family, and my relationships?

If the greed and misdeeds of humanity bulldoze my spirit, then I, too, fall victim, yet another statistic piled onto the growing mound of sorrows. If I give in, I prostrate my thoughts, emotions, and body to another's careless or carefully orchestrated tyranny. If I don't attempt resistance and establish a vital positioning of myself in this nightmarish scene, my heart may continue to beat, but I will have joined the family of the disempowered. I am unwilling to go the way of the hairy rhino and classify myself as another life extinguished, another hope mutilated, another vision scorched. I am unwilling to perpetuate violence and ignorance by turning it on myself. With life at stake, my preference is to assert my own vision with courage, steadiness, and compassion.

Substantial change takes effect when it happens on two levels. Work needs to take place on the cause, the situation outside ourselves. Equally important, work also needs to take place within.

The suggestion of combining outer and inner work may come as a surprise, yet the combination is a potent blend. The careful evolution of one can and does influence the many.

The state of our inner being directly influences our behavior, which in turn influences how people respond to us. A rant over the phone results in a hang-up, whereas a letter with facts, evidence, and a reasoned tone wins an audience. Knowing which action to take and for how long can be hard to determine. A torrent of activity provides temporary relief, but combining action with research, reflection, and planning leads to effective, sustainable outcomes. If we're not careful, well-meaning actions can take a hurtful turn. We overwork ourselves and lose patience. We confuse facts. We discredit actions. We blame and demonize the other. We throw bricks through windows, erase files, destroy equipment. Already upset by the balance of the world, inadvertently we contribute to the imbalance by not engaging ourselves more skillfully. We surrender our integrity and other values, justifying actions for "the cause."

Clarity of mind and intention coupled with equanimity function as a navigational star that helps us steer a course through the turbulent waters of confrontation. Every situation poses unique challenges. We need to know when to adapt to a situation and when to resist, how to use the materials at hand and when to seek additional resources, when to follow another's lead and when to assert a new vision. If we come from a place of disrespect, fueled by desperation or bitterness, or any other weakened place, we may become the type of person we abhor. It is vital that we engage the entirety of our being and extend ourselves like a firm yet receptive handshake. It is vital that we not perpetuate violence to our selves—or any self. It is vital that we clarify our intentions and ground our actions in compassion and respect for all life.

Rather than allow feelings to jerk us around, acknowledge and channel feelings into appropriate actions. Rather than *react* to the injustice, *respond* with consideration for the ramifications of our actions. Reactivity draws from unconscious, knee-jerk behaviors; responsiveness calls forth a well-planned approach. The actions taken will vary from person to person and, perhaps, month to month. Ask pivotal questions, perhaps daily: How can I engage my time, skills, and energy effectively? How can I sustain myself in peace, power, and compassion? What can I do to embody and instigate change?

As sea levels rise and suffering increases, we need to learn new ways to take better care of the body of earth and the body of being. Working on inner and outer realms requires both subtle and radical actions. What values and habits need dismantling? Which ones need cultivating?

What if, for instance, we shift status items from large houses to small; we prize collecting of good will over accumulating material goods; we collaborate on projects and ideas for win-win outcomes; and we appreciate art making and introspection over ceaseless industry? What if we measure the wealth of a country not by its Gross National Product, but by following Bhutan's lead on Gross National Happiness? What vision for the present and future nourishes life?

In *Coming Back to Life,* authors Joanna Macy and Molly Young Brown recognize three categories of activity: holding actions such as protests and letter-writing campaigns that slow or halt further harm; studying structural causes and developing alternative institutions such as in education and economics; and shifting our perspective cognitively and spiritually by engaging in creative and meditative practices like hiking and writing.

How is hiking to the waterfall helpful to social action? Time away from routine reduces stress. The physical activity of scrambling over rocks and crossing the stream provides a visceral experience of welcome balance that refreshes the spirit. Water tumbling over the cliff contrasts with our own halted momentum and reinforces the value of a fluid self. The journey positions us among the many forms of life together sharing ground and sky, sun and moon. We return to our jobs and homes restored.

Creative work similarly revives the spirit. In the pursuit of crafting expression, we interact with the raw material of our thoughts, emotions, and impressions. We actively participate in the flow of the world as it unfolds. We recognize a constant dialogue, an interdependency that takes place both behind the scenes of our awareness and on its central stage.

A similar interdependency transpires during practices such as meditation and yoga. Visceral, unmediated activities reveal the degree of connection between our mind and body, and between our self and our surroundings. These practices reinforce attentive witnessing to the emergence of sensations, thoughts, emotions, and impressions and how we engage with the world. Habits and opportunities come into focus and point toward the possibility of change. We see how the world "out there" filters into and reflects our inner world; conversely, our inner world impacts the outer world. The two intimately twine.

Attentive witnessing during a political demonstration is nothing less than a transformative act. Hiking or writing or meditating with an awareness of how the personal is political is equally transformative. All connect us to the edge of life in its moment of unfolding and position us as an active participant. Actions carried out with vigilance root us in ourselves, affecting, too, those who meet us or encounter our work. These actions bring us one step closer to a more sustainable world.

Setting an intention to bring about a more peaceful world helps us recognize opportunities and take them one step at a time. The collection of small and large steps makes a difference; taking no step at all contributes to failure.

Another recent article from Malaysia reports that the hairy rhino is not extinct, though its numbers are precariously meager, with only about thirty animals found on Sabah, an island of Malaysia. In response, a group of people formed a conservation organization devoted exclusively to the survival of these rhinos.

In what ways do you dwell within the body of your self and the body of this earth? If we take no action, the hairy rhino succumbs to our misdeeds. One gesture or a series of small steps contributes significantly to its recovery and to the balance of our world.

⤳

Writer, dancer, and university professor Cheryl Pallant is the author of several poetry books—*Morphs,* her most recent, as well as three chapbooks—and a nonfiction book on dance. Her short stories, articles, essays, and reviews have appeared in several anthologies and in numerous print and online journals. She has lived and taught at universities in Malaysia, South Korea, Hong Kong, and the U.S. Visit her online at www.cherylpallant.com

The World Doesn't Need to Be Saved

BYRON KATIE

I have looked down the barrel of a gun pointed at me, and never for an instant was I afraid. Fear is the story of a future. How could I know that he would pull the trigger? How can I know that an environmental catastrophe will happen or, if it does happen, that it will be a bad thing for me, for you, for the planet? Once you understand this, and begin to live in reality, not in your unquestioned thoughts about reality, life becomes fearless, loving, and filled with adventure and gratitude, whatever the nonexistent future may bring.

Fear is not possible when you've questioned your mind; it can be experienced only when the mind projects a story into the future. The story of an unquestioned past is what we continue to project as a future. If we weren't attached to the story of a past, we would notice that we're already living in the future, and that it's always now, and now is always good.

The war with God—which is another name for reality—always sees catastrophes looming, whether these are planetary or personal. It's a very painful way to live. But when you question your mind, thoughts flow in and out and don't cause any stress, because you no longer believe them. And you instantly realize that their opposites could be just as true, or even truer. Reality shows you, in that peace of mind, that there are no problems, only solutions. You know, to your very depths, that whatever happens is what should be happening and you know what to do. If I lose my life or my children or the whole planet, I lose what wasn't mine in the first place. It's a good thing. Either that, or God is a sadist, and that's not my experience.

In the meantime, I go about my business as if there were no life and no death (and there isn't). My home is powered by the sun, my Segway is powered by my home, the car I drive is a Prius, I'm careful about recycling, I vote for people who say they are concerned about global warming and have a record to prove it, I give money to environmental causes. I am fearless, worry-free, and I do whatever makes sense to me for my good, which is for the good of the whole. "Get solar panels," the mind says, and there is no possible reason not to do it, since all thoughts have been tested by inquiry. "I can't afford to do it"? "I can't afford not to do it" was truer. The panels are installed, my electric bill was minus $23 this month, and at some point I will have put back all that I have used, and more. This will match my existence: all traces gone, a grateful life given back to what it came from.

I once worked with a large group of environmentalists. These were people who had committed their lives to saving the planet. They were living with a great deal of anxiety, even terror, they said—an enormous burden on their shoulders. But many of them had open minds and were willing to question the thoughts that were causing them so much stress. I helped them investigate thoughts such as "Something terrible is going to happen," "I need to save the planet," and "People should be more conscious." They discovered how these thoughts were driving them crazy, and how the thoughts have various opposites that might be just as true or truer.

After a few hours of intensive inquiry, I asked them to imagine the worst thing that could happen: a full-scale environmental disaster that wiped out humanity. They shared their fears and gave a lot of graphic details. Then I asked them to turn the thought around, to find the thought's exact opposite: "The *best* thing that could happen is a full-scale environmental disaster that wiped out humanity." I asked them each to give me three reasons why this statement could be as true as, or truer than, the original. And these brave people really were able to go there: "It might be good for some endangered species not to have people around." "It would be good for insects." "We wouldn't be pumping and mining the life blood out of the planet." "It would be good for the rain-forests." "Who knows what intelligent species would evolve if we were gone?" And there were many more ideas like this.

Inquiry is grace. It wakes up inside you, and it's alive, and there's no suffering that can stand against it. It will take you over, and then it doesn't matter what experiences life brings you, "good" or "bad." You find yourself opening your arms to the worst that can happen, because inquiry will continue to hold you, safely, sweetly, as reality does, through it all. Even the most radical problem becomes just a sweet, natural happening, an opportunity for your own self-realization. And when others are experiencing terror, you are the embodiment of clarity and compassion. You are the living example, the match for reality.

One of the things you discover when you question your mind is that the world doesn't need saving. It's already saved. What a relief! The most attractive thing about the Buddha was that he saved one person: himself. That's all he needed to save; when he saved himself, he saved the whole world. All his years of teaching—forty years of apparent compassion—were just the forward momentum of that one moment of insight.

I don't order God around. I don't presume to know whether life or death is better for me or for anyone I love. How can I know that? All I know is that God is everything and God is good. I call this the last story.

Reality is kind. Its nature is uninterrupted joy. When I woke up from the dream of Byron Katie, there was nothing left, and the nothing was benevolent. It's so benevolent that it wouldn't reappear, it wouldn't re-create itself. The worst thing could happen, the worst imagination of horror, the whole planet could be obliterated, and it would see that as grace, it would even celebrate, it would open its arms and sing "Hallelujah!" It's so clear, so in love with what is, that it might seem unkind, even inhuman. It cares totally, and it doesn't care at all, not one bit, not if all living creatures in the universe were obliterated in an instant. How could it react with anything less than joy? It's in love with what is, whatever form that may take.

As you begin to wake yourself up from your dreams of hell or purgatory, one by one by one, heaven begins to dawn on you in a way that the imagination can't comprehend. And then, as you continue to question what you believe, you realize that heaven, too, is just a beginning. There is something better than heaven. It's the eternal, meaningless, infinitely creative mind. It can't stop for time or space or even joy. It's so brilliant that it will shake what's left of you into the depths of all-consuming wonder.

If you have a problem with people or with the state of the world, I invite you to put your stressful thoughts on paper and inquire, and to do it for the love of truth, not in order to save the world. Is your thought true? Can you absolutely know that it's true? How do you react—what happens—when you believe that thought? Who would you be without it? Then turn the thought around: save your own world. Isn't that why you want to save the world in the first place? So that you can be happy? Well, skip the middleman, and be happy from here! You're it. You're the one. In this turnaround you remain active, but there's no fear in it, no internal war. So it ceases to be war trying to teach peace. War can't teach peace. Only peace can.

I don't try to change the world—not ever. The world changes by itself, and I'm a part of that change. The world changes through me, as the mind changes. I'm absolutely, totally, a lover of what is. When people ask me for help, I say yes, I teach them how to question their stressful thoughts, they begin to end their own suffering, and in that they begin to end the suffering of the world.

Violence teaches only violence. Stress teaches only stress. If you clean up your mental environment, we'll clean up our physical one much more quickly. That's how it works. And if you do that genuinely, without the violence of fear in your heart, without anger, without condemning corporations as the enemy, then people begin to notice. We begin to listen and notice that change through peace is possible. It has to begin with one person. If you're not the one, who is?

I'm open to all that the mind brings, all that life brings. I have questioned my thinking, and I've discovered that ultimately it doesn't mean a thing. I shine internally with the joy of understanding. I know about suffering, and I know about joy, and I know who I am. Who I am is who you are, even before you have realized it. When there's no story, no past or future, nothing to worry about, nothing to do, nowhere to go, no one to be, it's all good.

❧

Byron Katie's simple yet powerful method of inquiry into the cause of all suffering is called The Work. Since 1986, she has introduced The Work to millions of people throughout the world. Eckhart Tolle says that The Work is "a great blessing for our planet" and *Time* magazine named Katie a "spiritual innovator for the new millennium." Her three best-selling books are *Loving What Is, I Need Your Love—Is That True?* and *A Thousand Names for Joy;* other books are *Question Your Thinking—Change the World, Who Would You Be Without Your Story?* and, for children, *Tiger-Tiger, Is It True?* Her Web site is www.thework.com.

What Keeps Me Alive:
Making It Real

CHAIA HELLER

I've been teaching and engaging in activism on issues of feminism, ecology, food, and agriculture for well over twenty years. Once in a while, I run into a face I recognize from the "old days." People hailing from the 1970s to the 1990s—times when there were feisty, in-your-face kind of grassroots activists working toward lofty goals, like ending patriarchy or stopping nuclear power. Sometimes folks ask me what I'm doing.

"I'm still teaching and writing about ecology and revolution, taking action when I can," I'll say, always slightly surprised to hear the words plop from my mouth like a series of plum stones. "You're still at it?" My interlocutor will ask, his or her voice a cocktail of sympathy and disbelief.

Truthfully, the disbelief is mine. I find it hard to comprehend how folks in their middle years can drop the proverbial revolutionary ball. "How do you stay inspired or optimistic?" my young students ask each year, amazed to see a relic like me still maundering on about the wonders of a potentially utopian, ecological, and directly democratic society.

And then I tell them: I had good teachers.

When I was twenty, I stumbled into the Institute for Social Ecology, a very small pond where a few big green fish swam within its humble perimeters. There, I met lifetime activists such as Murray and Bea Bookchin, Dave Dellinger, and Grace Paley, to name but a few of the zestiest. While Dellinger and Paley were mainly known as anti-militarist activists, Bookchin was involved in the ecology movement since its inception in the 1960s.

While living modestly, these peoples' lives were filled with sumptuous dreams they worked to transform into reality. While they certainly suffered from bouts of frustration and dismay over the years, none of these greats surrendered to what we call today "political burn-out"—or worse, just plain jadedness.

"How do you do it?" I used to hear myself ask my own teacher, Murray Bookchin, who taught and inspired me during my formative years. "How do you stay so furiously engaged?" I'd ask Murray, whose eyes refracted the light of the fires raging during the Parisian Sections—a righteous historical struggle where everyday people fought for the right to live in a decentralized directly democratic society.

The fact that they lost was immaterial. The fact that they imagined such a world and fought for it: now *that* was something.

"How do I do it?" Murray would ask, roosting deep into an old overstuffed chair in Bea Bookchin's living room where Murray used to lead study groups on social ecology during the 1980s. "Once you've tasted the flavor of freedom on your tongue, how can you not yearn for another taste?" Murray said, chuckling.

Even back then, I knew it wasn't that simple. As I matured a bit, I learned that this level of commitment had to do with the "motor" if you will, that drives the political theory and activism of the "long-time doers."

"It's about ethical versus instrumental reason," Murray said to me, so many times over the years. This discussion buoyed me up as I tried to cultivate my own revolutionary focus and morale. "If you are driven by instrumental reason alone," Murray said, "you will be guided by logic of efficiency and pragmatism. You will be forever running toward the lesser of two evils rather than pursuing our own dreams. If you are driven instead by ethical reason, then you will be guided by logic of what *ought* to be, what is just and humanistic—as well as simply "doable."

During my life, I have noticed that environmental activists remain active only when driven by ethical reason. The long-rangers have a gleaming golden carrot of justice dangling before them. This carrot is palpable and irresistible; they cannot ignore it as they go about their lives.

However, if they yield to logic of instrumentalism—pervasive in an age of bureaucracy and neo-liberalism—then our eyes will be locked forever on the grim and dank "bottom line," calculating the "feasibility" or "productivity" of our efforts.

If environmental activists evaluate their work in terms of immediate efficacy and pragmatic "do-ableness," they often collapse after five to ten years (sometimes far fewer) under the weight of abject disappointment. They resent themselves, their movements, and the world, for not changing fast enough.

The old timers I used to know—the kind of person I pray to become—were driven by ethical reason. They believed that what is utopian is just as real as the world in which we live. To be a little fancy, I'll refer to Hegel's truism about the worlds of the real and the actual. For Hegel, the actual world is the world now, as it is. It represents all of the "actualities" that swirl around us like the dizzying rings of Saturn. The world of the real, on the other hand, is one sculpted by justice and ethics. The idea of small-scaled, confederated, ecological, and directly democratic societies is a reality worth fighting for.

It's real to me.

Murray Bookchin was the kind of visionary who saw the "shining city" of freedom. He had the rare ability to articulate it, making it real for those around him. This shining city stands just at the glittering edge of a pink silvery sunrise. Sometimes I see its phosphorescence in a puddle of moonlight right beneath my feet on an otherwise rainy and dismal night.

Idealism, utopianism—grounded in a well-articulated (and experimentally oriented) theory, drive me to keep walking toward the shining city—the city, town, or village in which each person and living creature has the ability to fulfill their potential for greatness—whatever that potential may be.

In his last weeks of life, Murray lay in bed, composing short pieces, hoping his strength would kick in again and he could return to a life's work that did not end until he drew his last breath. His revolutionary vision—one anchored in knowledge of revolutionary history, philosophy, and a love of humanity—kept him not only alive, but completely passionately engaged for eighty-five years. Had he lived to be one hundred, it would have lasted just that much longer.

I pray each day to do right by my mentor—to do right by all of the lifelong revolutionaries who kept their eyes on the shining city, even when it was clouded by moments of self-doubt, discouragement, and fear.

Before I do yoga each day, I light three candles. The first candle is for *expansiveness,* reminding me to reach toward the challenges the world presents each day. The second candle stands for *acceptance,* beseeching me to accept what cannot be changed while also accepting the responsibility to mend whatever I can. The third candle is for *gratitude.* Gratitude is an eternal oil, burning ceaselessly in the activist's lantern without ever drying into an oily dark crown of emptiness.

At the end of my yoga practice, I kneel before these candles, erasing each flame with my breath. I thank my teachers, living and gone, for handing down their precious mantles of wisdom, strength, and compassion. I then rise again to my feet, striding out of my temporary sanctuary. I saunter back into the world to give it another glorious try.

❧

Chaia Heller teaches anthropology and gender studies at Mount Holyoke College and taught at the Institute for Social Ecology in Vermont for over two decades. Chaia has been involved in the ecology, feminist, anarchist, and global justice movements an activist, educator, and writer. Chaia received her PhD in anthropology from the University of Massachusetts, Amherst. Her new book, *Post-Industrial Peasants,* is forthcoming from Duke University Press. It explores the role of radical French peasants in the international controversy surrounding genetically modified organisms. She is a recipient of fellowships from the National Science Foundation and the Centre de Societe de l'Innovation. Chaia teaches gender studies at Mount Holyoke College and is also the author of *Ecology of Everyday Life: Rethinking the Desire for Nature* (Black Rose Books).

In the Climate Era
the Personal Is Political

TZEPORAH BERMAN

When I was fourteen my parents died and I remember wondering how the world could continue. Seeing people go about their daily lives when I felt as though a hole had been ripped through my heart left me astonished and alone. It took weeks before I could function in society, months before I stopped feeling like I was sleep walking and years before the nightmares stopped and I began to feel whole again.

The day I set my despair free was the day I married a sense of purpose with my loss. I was speaking at a rally to protect old growth forests in British Columbia on the steps of the legislature in Victoria. I remember being horrified by the rate and extent of clear cutting on Vancouver Island and feeling a familiar deep sense of loss. Closing my eyes to over a thousand people standing on the legislature lawn I grabbed the microphone and gave it everything I had—lamenting the loss of these great thousand year old trees and decrying the tragedy of the last of the wild being destroyed to make phone books and toilet paper. The roar of the crowd was deafening and when I stepped off the stage I was shaking. Standing by the steps was an elderly couple openly weeping. They stepped forward and introduced themselves as old friends of my parents and told me that they knew my mom and dad well and that they would be very proud. That day I realized that beyond the pain, fear, and anger was a place that sparkled with purpose, pride and honor. Over the next decade I would remember that moment over and over again. I would remember how I could channel that despair and anger into action and how good it feels to know that you are contributing to something that will have an impact beyond your own life time.

This memory has come back to me over the last year as I've poured over the new United Nations reports that now say unequivocally that a two degree warming of the planet is now unavoidable due to human activity and that this will lead to "catastrophic climate change"—floods, droughts, food shortages, and species extinctions. Whether it comes from the loss of a loved one or from the knowledge of the threat of global warming, despair is a powerful place to step forward from. There are times when it feels too big, too overwhelming and I feel too small and insignificant. But everyday I look at my children and know that I need to find ways to contribute to raising awareness and finding solutions. I need to continue to find my voice and allow myself to feel the pain of despair to motivate a new strong commitment and voice for change.

I think the answer to how we live our life when we are looking down the barrel of environmental catastrophe lies in recognizing and spending some time in despair, so that we don't end up denying or avoiding reality. Global warming will not be solved by changing a light bulb. Species extinction will not be slowed down by creating a park. The magnitude of the global environmental collapse we find ourselves in is unprecedented. Last year the Millennium Ecosystem Reports prepared by over a thousand scientists globally concluded, "Human activity is putting such a strain on the natural functions of the earth that its ability to maintain human well-being is now in question." We are living in a time of crisis. A time that requires us to step up to our fullest potential. To live our lives intentionally and to use every moment, every purchase, every conversation as a vehicle for change. As in wartime, we are being called upon to make sacrifices, to change our lives for the greater good.

My family has discovered that these lifestyle changes actually lead to a fuller and more engaged life. We take walks and ride our bikes together instead of driving. Last winter we hand rolled beeswax candles and canned our own jam for holiday presents instead of consuming goods that traveled long distances and were made from finite resources. That said I know that lifestyle changes alone are not enough. If we are going to mitigate global warming it requires significant shifts in how our society functions, how goods are made and transported, how major industries operate.

These large-scale changes require government legislation to protect the very air we breathe, the water we drink, and the climate that sustains all of us. We need to live our lives looking for the places that we can engage. Protesting decisions that don't reflect the urgency of the changes that we need and organizing within our communities and beyond to explore the tough transition to a carbon neutral society. At the organization I founded, ForestEthics, we are expanding our work to address oil and gas issues, we have stepped up our efforts to expose corporations and governments who are claiming to be "green" without changing their practices. We have increased our outreach and list-building to build a powerbase that can engage with governments through letter writing and calling campaigns. And most importantly we have been creating 'untraditional alliances' by working with some of the major corporations who are willing to make change. I offer you these examples not because they may be right for you but to offer a window into how exciting it is to let go of old ideas and alliances and work with a group of like minded people to make change.

In the climate era, "the personal is political." The decisions that we make in our own lives, what car to drive or what food to buy—have significant consequences. Perhaps more importantly is how we chose to spend our time and with whom. Whether we choose to write a letter to our elected official, engage in our community or with an environmental organization, or to protest. These decisions are critical factors in whether those in power will feel the will and the support to make the hard decisions to reduce greenhouse gas emissions and conserve what's left of the wild. We need to be bold, to be creative, to step up and be willing to take risks to make change. In the end it is clear to me that we need to live every moment remembering that we are responsible today not only for what we do but for what we don't do.

≫

Tzeporah Berman is the co-founder of ForestEthics and the co-director of Greenpeace International's Global Climate and Energy Program. Tzeporah was one of the experts interviewed in Leonardo DiCaprio's environmental documentary *The 11th Hour,* was one of six Canadian nominees for the Schwab Social Entrepreneur of the Year Award, and has been profiled as one of "50 Visionaries Changing the World" in *Utne Reader.* She blogs on www.zerocarboncanada.ca and can be found on Twitter at www.twitter/tzeporah.

Coping with
New Realities

LINDA BUZZELL

Like Neo in the film *The Matrix,* perhaps you're waking up to the true extent of the challenges we humans face at this point in history. Some call this the "oh shit!" moment. Now what do we do? How can we enjoy life on a day-to-day basis in light of what we are beginning to understand about the truly terrifying collective disasters and challenges headed our way?

I'd like to share some of the advice that I give my psychotherapy and ecotherapy clients who are in the process of experiencing what Dr. Sarah Anne Edwards and I have dubbed "the waking-up syndrome":

Be gentle with yourself. This is a lot to take in. Most of what we thought we knew about living is turning out not to be true. We're bombarded with unwelcome news on a daily basis. Global warming. Fossil-fuel depletion. Animal extinctions. Resource wars. Economic instability. Poisoning of our environment. Corporate crimes. Renewed nuclear threats. Many of us are now worrying about the damage we do with every mile we drive, bite we eat, and imported jacket we wear. Old dreams and illusions shatter. Eco-guilt and eco-anxiety rise. We go in and out of denial, just as Neo does in *The Matrix.* Sometimes we just want to block it all out and party. Sometimes we're fired up to take action. Sometimes we're just depressed. It's all part of the process, so "easy does it."

Be patient with your loved ones and co-workers. Once you're awake, it's tempting to become a modern-day Paul Revere, obsessed with the news, crying out to one and all about what's headed our way. I've seen a number of friends and clients succumb to this, neglecting the rest of their lives and their loved ones and even their own safety as they dig for the truth and spread the word to others. A few divorces have resulted. It's important to face some home truths: your kids may not want to give up McDonald's; your spouse may not share your concerns for the fate of the planet or your enthusiasm for new solutions; your co-workers may still prefer paper and plastic to ceramic cups. Just try to "be the change you want to see" and don't attempt to convert everyone you know. Many more folks will wake up or get involved as time goes on and conditions worsen, but it's important to let them do so in their own way and time.

Don't linger too long in the negativity and the problems without getting into positive action. Eco-philosopher Joanna Macy suggests that those of us who are concerned about what's happening on our planet will be able to protect our sanity and balance our lives if we engage in three types of activities simultaneously. Here's my simplification of what she recommends:

1. Resist what's wrong. Get involved in (or start) a group that speaks up against one or two things that are destructive. But don't try to save the whole world all at once.
2. Build what's right. Create or be the solution(s). Craft lifeboats, put in a garden, make some of your own clothes, start a neighborhood exchange.
3. Raise the level of your awareness and consciousness. Find a spiritual practice, take a permaculture class, learn about systems theory.

I see this as a sort of "Chinese menu" approach to mental health in light of our present circumstances. Pick one activity from category #1, one from #2, and one from #3.

The Call to Action

Macy's three-part prescription is very wise, encouraging us to find a balance in our lives so we have the resilience to deal with challenges. It can be dangerous to both you and your loved ones if you spend too much time in the war against the bad guys (Category One). It's far too easy to burn out on political and social activism if you're not also involved in more positive activities (Category Two and Category Three).

But it can also be counterproductive to avoid the call to battle. In fact, the only cure for the tumultuous emotions that arise as one wakes up to the challenges is to get into constructive action as quickly as possible. Hanging out too long in denial, guilt, or eco-anxiety is downright unhealthy, both for you and for the human future. Most of us who are aware of what's really happening right now won't sleep very well at night if we don't get involved in some way, doing our bit by trying to put a stop to at least some of the bad stuff, whether it's environmental, social, or political. One's own consumption is a great place to start.

The trick is to balance that "warrior" activity with at least one activity from Category Two and one from Category Three.

Be the Change

Category Two offers a lot of joyful, fun opportunities to get involved with creating and building positive, hopeful lifeboats and solutions. Work on building a local school garden. Create new bike paths. Plant fruit trees. Take a class on grey water plumbing and water harvesting. Learn how to do permaculture design. Move into an eco-village or co-housing. Do a green remodel. Join a Simplicity Circle.

Bottom line: start building a local, sustainable lifestyle for you, your family, and your local community.

Find Your Inner Serenity

As we do the work to create alternatives, Category Three invites us to develop some sort of spiritual practice to keep our equanimity as we take a stand against those who actively destroy life. This practice can be as simple as walking in nature or taking a yoga class—or as complex as learning a sophisticated meditation technique, studying "Creation Spirituality" and the Universe Story, or immersing ourselves in the new and old wisdom traditions from our own or other cultures. It can also involve deep intellectual exploration of Earth's biological systems and the history of our species and the many other species sharing this planet. By engaging in Category Three activities, we seek a long view that insulates us from the immediate ups and downs of daily life.

But don't get lost in that long view. Spirituality can be so personally satisfying that it can lure us into spending most of our time exclusively in Category Three, as some meditators, yogis, and New Age folk have done. It's important to realize that by focusing only on our own spiritual life or yoga practice to the exclusion of growing real-world threats, we can inadvertently become part of the problem rather than doing our part to find solutions.

The core message here is that it's not a good idea to hang out exclusively in any one of these modes to the exclusion of the others. Even if you're working hard building, say, a permaculture ecovillage (Category Two), life goes better if you include some Category One and Category Three activities in your day as well.

Take practical steps toward making what author James Howard Kunstler calls "other arrangements" in your daily life—and get support from like-minded people as you do so. My husband and I joined a Simplicity Circle eleven years ago and have been very grateful for the group's ongoing encouragement as we take slow and incremental steps away from the consumerist, materialist Western lifestyle and learn new survival skills. What has worked for us: getting out of debt, cutting up our credit cards, de-cluttering our house, spending less, saving more, shopping as locally as we can, downsizing our lifestyle, limiting the time we stare at screens, working fewer hours, taking a permaculture design course, putting in an organic garden and backyard food forest, learning how to cook locally and seasonally from our garden and the farmers' market, working with neighbors to share the bounty, and enjoying the good (sustainable) life.

Move your work life toward sustainability. Now that you are awake to the import of the historical moment, it may be difficult to continue working in a profession or job that is either part of the problem or doesn't speak to your soul. There are a number of possible moves here:

Transform your own job and profession. Perhaps you can continue to do what you do but more mindfully and compassionately . . . or perhaps you become active in changing things in your field. For example, if you're a physician, you might explore alternative or integrative medicine, a rapidly growing field. If you're a teacher, you might want to learn more about Waldorf education or the home-schooling movement. If you're a carpenter, you might consider salvaging local trees that have been cut down and milling them yourself to create local lumber.

Continue in your present career, but use it as a "day job" to support your true work in sustainability. Your current job may pay for further education, or let you save up money to build a new sustainable business, or support you while you volunteer in an activist organization. A friend of ours does window-washing as a day job that allows him to engage in his true vocation as a cutting-edge publisher/editor of an alternative magazine and a political/environmental film exhibitor. Another friend does elder-care as a day job while he's preparing to transition into permaculture design as his "next career."

Transition now into a new, more sustainable profession that calls strongly to you. Perhaps you'll return to school full-time, or maybe you'll start a new business offering goods and services that will be needed in the sustainable future and emerging green economy. Sometimes you just need to take the leap.

The result of all these changes is a life full of joy, new friends, good food, and lots of time spent in nature, with loved ones and on deeply meaningful activities.

The hour is late, and a growing number of us daily become more aware of the necessity of transitioning quickly to sustainable society and building local and global communities that truly support their members. The sooner we begin to reshape our own lives to fit the new times, the happier and healthier we will all be.

RESOURCES

Linda Buzzell and Craig Chalquist, editors. 2009. *Ecotherapy: Healing with Nature in Mind.* San Francisco, CA: Sierra Club Books. www.ecotherapyheals.com

Sarah Anne Edwards and Linda Buzzell. 2009. "The Waking-Up Syndrome," from *Ecotherapy: Healing with Nature in Mind.* San Francisco, CA: Sierra Club Books.

Rob Hopkins. 2008. *The Transition Handbook: From oil dependency to local resilience.* Totnes, Devon: Green Books. www.transitionculture.org

Joanna Macy. 2007. *World as Lover, World as Self: Courage for Global Justice and Ecological Renewal* (Berkeley, CA: Parallax Press).

☙

Psychotherapist Linda Buzzell is the editor with Craig Chalquist of *Ecotherapy: Healing with Nature in Mind* (Sierra Club Books, 2009), an anthology of writings on healing the human-nature relationship that includes essays by Joanna Macy, Andy Fisher, Richard Louv, Ralph Metzner, Bill McKibben, and Richard Heinberg and a Foreword by David Orr. She is the founder of the International Association for Ecotherapy and editor of its quarterly publication *Ecotherapy News* (Online at www.ecotherapyheals.com). She is an official blogger about ecotherapy at online newspaper The Huffington Post: www.huffingtonpost.com/linda-buzzell.

HOPE BENEATH OUR FEET

CHAPTER SEVEN

Meditations on Living
in These Times

Eden Is
a Conversation

BARRY LOPEZ

In May of 2006, more than five hundred people from forty countries gathered in Ubud, Bali, Indonesia, to participate in "Quest for Global Healing," a determined effort to address pressing social and economic problems around the world. Among the speakers were Archbishop Desmond Tutu; two other Nobel peace laureates, Betty Williams and Jody Williams; Gus Dur, the leader of Indonesia's forty-five million Muslims; Fatima Gailani, head of Afghanistan Red Crescent; Thai political activist Chaiwat Thirapantu; Bhutan's minister of Labour and Human Resources, Lyonpo Ugyen Tshering; and Barry Lopez, who delivered the following as his closing talk.

A few days ago I visited Tirta Empul, a temple just north of here, the site of Bali's holiest spring. I spent almost an hour gazing into the large rectangular basin of waist-deep water, transparent as a polished windowpane. The water rose from a spring obscured on the sandy bottom by water plants, and flowed away south from the temple's lower courtyard, south and father south through pools and basins on the temple grounds, then a series of canals, on southward to I don't know where.

The interior courtyard, late in the day, was nearly empty, quiet as the surface of the moon. The only sound came from the swooping flights of swallows feeding off the surface of the pool—the rush of air over their bodies, the click of ligaments in their wings responding to their swift and acrobatic movements.

I needed this interlude in our stimulating conversations, and have to think others of you have sought out similar nearby spots on the Balinese landscape. Entering these temples—perhaps you felt something similar—

209

I felt a kind of divestiture, a stripping away, an opening and vulnerability in the presence of Hindu spirituality. It made little difference that this was not my chosen faith. This seemed incidental in the face of what was apparent.

We have been in a kind of temple of our own making over the last five days, doing our best to elevate and embrace a protracted conversation. Now, the hours of leave-taking have come—a last look at the limpid holy water, its language shimmering in the ancient stone basin.

Driving back to Ubud through Bali's handmade landscape, a countryside of supplication and spiritual courtesy, where one sees endless signs of a studious attention to elements of enchantment in the place, it occurred to me that leave-taking at a temple is an undertaking just as important as entering such a place. You enter, aware of the centuries of people who've come like yourself—hopeful, scared, humble, desirous. You leave refreshed, rededicated—hopeful, scared, humble, desirous.

We are leaving the temple now, and carrying, each of us, a special kind of determination, a desire to do good beyond the self. And we are carrying along with this the spiritual resonance of Bali, a place some call a kind of Eden.

But Eden, we should be at pains to point out, is not a place. Eden is a conversation. It is the conversation of the human with the Divine. And it is the reverberations of that conversation that create a sense of place. It is not a thing, Eden, but a pattern of relationships, made visible in conversation. To live in Eden is to live in the midst of good relations, of just relations scrupulously attended to, imaginatively maintained through time. Altogether we call this beauty.

We have heard from some remarkable people, people in remarkable service to humanity and place, people pursuing good relations, just relations, reverent relations all over the world. Peru, South Africa, Ireland, California, Thailand. We've been urged to join in.

We cannot, of course, save the world, because we do not have authority over its parts. We can serve the world though. That is everyone's calling, to lead a life that helps.

We have heard a surprising, wise, and inspiring description of the pursuit of happiness, and it has filled our discussions of pursuing virtuous relations in our lives, occupied our conversations as naturally as the water at Tirta Empul fits that basin. Happiness is an awareness of the presence of good relationships, harmonious antiphonies, reciprocities in which you are included, in which your participation is essential, and for which you are glad to be held accountable. Happiness grows out of the practice of virtuous behavior, out of service to the Divine as it becomes apparent to us in humanity, in the earth and its creatures.

We have spoken thoughtfully of action to heal human damage all over the world, but it is enough, really, to enter into, to craft, beautiful conversation to know that our time in Bali was well spent. If we have understood in our days together the need for good conversation, for generous, attentive, courteous, and respectful exchanges, every strategy for change we can imagine will have a good foundation.

We cannot save things. Things pass away. We can only attend to relationships, to the relationships between things. It is here that we see the most beautiful images we are capable of apprehending or imagining—the relationship between a mother and a child, the racket of sunlight on pooling water, a bird alighting on a limb.

Conversations are efforts toward good relations. They are an elementary form of reciprocity. They are the exercise of our love for each other. They are the enemies of our loneliness, our doubt, our anxiety, our tendencies to abdicate. To continue to be in good conversation over our enormous and terrifying problems is to be calling out to each other in the night. If we attend with imagination and devotion to our conversations, we will find what we need; and someone among us will act—it does not matter whom—and we will survive.

We need to thank our ancestors, who knew trouble was coming and whose prayers have brought each of us to Bali to meet, to draw wisdom and strength and renewal from each other. We need, each according to his or her gifts and by his or her own lights, to be the servants of beauty. We need to prefer being in love to being in power. We need to know that as we have met and now come to a close here in Bali, others with hearts like ours have been gathered—in Islamabad and Chengdu, in Winnipeg and San Miguel de Allende and Santiago, in Sapporo and Irkutsk and Sydney. In the villages of Alaska and India, of Nigeria and Oceania they are embracing, affirming diversity and solidarity, making vows and stepping off like blazing torches into the thousand nights that lie ahead.

Just like us.

Cherish each other. Travel in beauty. Our lives depend on it.

❧

Barry Lopez is the author of thirteen books and numerous articles and short stories that have appeared in the United States and abroad. He received the National Book Award for *Arctic Dreams* and is a recipient of the Award in Literature from the American Academy of Arts and Letters; the John Hay, John Burroughs, and Christopher medals; and awards from the Academy of Television Arts and Science and other institutions. His short story collections include *Resistance* and *Light Action in the Caribbean,* and his essays are collected in *Crossing Open Ground* and *About This Life.* For more, see www.barrylopez.com.

Fostering Light
in Dark Times

VIVIENNE SIMON

I have always been fascinated by change and transformation. For most of my life I've been an activist in international environmental and human rights efforts, and for much of that time, immersed in various kinds of personal growth work as well. Despite having spent a lifetime engaging opportunities to foster growth and change, I now find the world has become a real-time living laboratory for observing and participating in transformation on a scale beyond anything I've known. We have reached a tipping point, and the natural and human-made structures are imploding and transforming before my eyes. For wisdom and guidance facing these troubling and challenging times, I reflect on what I've learned over many years of activism, and even more so, on my longtime Buddhist practices.

Hours of sitting on a meditation pillow, and the wisdom of many teachers, have taught me that the way to face everything—regardless of magnitude—is with mindfulness. By staying present for the truth of what is happening—without either turning away in fear or clinging to hopes that somehow this is all going to magically turn around or prove to be a false alarm—I'm able to tap into the deep animal in me that loves this planet and wants to protect my home, my son, and all I hold dear here. My love for that animal guides me in finding ways to be of use while also continuing to find pleasure in being alive.

I am mindful that no one person can save the world, and each of us can do something important to make a difference.

I am mindful that the changes in the planet are happening faster and more ferociously than scientists predicted. There's no turning back the tide of change that is in motion, and we need to open our hearts and minds fully to what is happening in order to engage in meaningful and effective work.

I am mindful that there will be great loss of life and enormous destruction to ecological systems, and that families and communities will be torn apart. There will be unimaginably huge numbers of people we cannot save from harm, and we can save our collective humanity by choosing to share the resources of the planet equitably rather than each hoarding what we can for ourselves.

I am mindful that love is the most powerful force for change and that keeping our hearts open, no matter how challenging things become, will keep us connected and capable of meeting the trying situations ahead.

I am mindful that I'm carrying tremendous despair and that I need to continuously allow that despair to crack me open and teach me greater courage and compassion.

I am mindful that there are many who live in denial and will resist change until they are forced by personal circumstance to respond, and they will come around in their own time. My job is to not get distracted by how others behave but rather to find my path of contribution.

I am mindful that all of existence is a great mystery, and that in the grand scheme of things stars, planets, galaxies, and universes are born and die, and it is the natural order of things. And the mystery is so much larger than I can understand in this human form. I, too, was born and will die and return to that mystery, and things that are incomprehensible to me now will make sense from another perspective.

I am mindful that the arts, friendship, laughter, ocean waves, birds in flight, the sunset, a beautiful song, and so much more sustain me as much as breathing, and they are what inspire me and make life worth living and protecting. Acting out of anger, giving in to despair, or burning out make me ineffective and isolated, which only compounds the problems.

I am mindful that finding like-minded, like-spirited comrades is essential to keeping up my spirits, fostering creativity, expressing love, and keeping me going.

HOPE BENEATH OUR FEET

I am mindful that we are in uncharted territory and that those who claim the mantle of leadership have failed us miserably. It is necessary to turn from all conventional wisdom and open to completely new paradigms in human relations, both to each other and to the natural world. We need to let go of economic systems that are based on winners and losers and create relationships grounded in the sacredness of life and creation. And we are all equally well equipped to be leaders, speak our truths, and offer insight and direction.

I am mindful that the earth has gone through evolutionary upheavals before and has always found her way back to being a hospitable place for life. And every time she came back renewed she was vastly changed, and new life forms and new natural formations appeared and thrived for thousands of years.

I am mindful that there are forces and spirits in the unseen realms that are here with us, available and able to help if we invite their assistance and honor it when it arrives. Regularly engaging in ceremony and prayer keep us connected to that power, and these rituals need to be brought back more broadly into our individual lives and communities.

I am mindful of the importance of self care, of living a healthy, balanced life, of connecting with heart and authenticity, of caring for my body, of speaking the truth, of dealing with hard emotions, and of spiritual practice—all vital to staying healthy so I can be involved for the long haul.

I am mindful that my many gifts, talents, and skills are of huge value and are greatly needed, and it doesn't serve anyone or anything for me to play small. I am an experienced elder in successfully bringing together personal and institutional change; in holding light through dark times; in physical and emotional healing; in providing leadership and in mentoring leaders; in asking probing questions and listening with openness and attention to answers wherever they come from. It is my work to pass on the wisdom, skills, and insights that my years on the front lines have sharpened, and to allow the forms that it takes to organically evolve.

I am mindful of an intelligence that permeates everything, that is in the core of every cell that has existed since the birth of the universe. It is also in the fires and the floodwaters and the dying coral reefs and the carbon that is suffocating us, and it is in me as powerfully as it is everywhere else. I can tap into that for good. And so can we all.

❦

Vivienne Simon, JD, CPCC, has devoted her life to fostering a just and sustainable world. She works on Amazonian rainforest protection, international anti-nuclear campaigns, famine relief in Africa, women's empowerment programs, consciousness research at Harvard, and the development of the University of the Wild. Her coaching, trainings, and writings help evolving leaders to respond to the unique challenges and opportunities of the twenty-first century with an engaged co-creativity. Vivienne's work draws on Eastern and Western practices and teachings and promotes the sacredness of all life. Her Web site is www.vivsimon.com.

From Mourning into Daybreak

NINA SIMONS

We've forgotten how to mourn. Lost the art of grieving.
 No one keens anymore.
Women in Greek tragedies knew how, but these days, we medicate.
We veer away from the depths,
 and so we rarely even see the peaks.
Now, we have become an unfinished circle—
 a culture caught recycling our wounds
because we don't ever acknowledge the pain, grieve our losses,
 complete the cycle of mourning.

How will we ever see daybreak without mourning?
If we don't feel what hurts, surrender to its demands,
 speak the wound, how can we really begin to heal?

I understand young people's prolific piercings now,
 black rings and claw-like ornaments jutting through their skin.
Wanting to wear some mark of realness, courage,
 a willingness to feel pain.
It's a modern-day sun-dance, dancing to awaken the world.
Proof that you're not to be counted among the anesthetized,
 those among us who are lulled into false security,
who've chosen this way of shutting down,
this course of least resistance, this blithe consumer life.
Those who are lured by promises of "safety"
and seek more stuff to distract themselves from feeling.

I was pierced recently, feeling the sudden loss of a friend's best friend,
    knowing I had no words to offer her comfort.
I encouraged her to dive down deep, to immerse herself
    as far down into her grief as she felt drawn.
To sing for her beloved friend, to wail, paint, dream, carve,
    dance the sadness.
To let the loss impregnate her belly, fully, without hurrying
    the passage.

When my father died, I felt the rock I stood on suddenly gone,
    my self in free fall.
I was warned that it might take a year for me to heal.
It was at least that, and it was longer.
I was grateful for the crystalline expanse of time I permitted myself,
    mindful of the warning.

I entered expanded elastic months of feeling transparent,
    of squinting at the striking brightness of colors, line, and light.
I oscillated between emptiness and attunement—
    the tenderness of my tears for him always only a breath away.

I was appalled to discover our illiteracy toward death,
envying the Japanese tradition of wearing a black armband
    for a year following the loss of a loved one,
so that everyone knows not to treat you in the usual way.

The tattoo I got this summer—
    permanent art marking my impermanent body.
A gift to myself—the pain a strange kind of prayer,
    making good on my promise to remember
    the loving hand of the feminine
always over my heart, having my back.

A huge crow swoops to meet me, his large beak stuttering open.
He croaks his hello, frog-like.
He must be drawn by the bones the dogs have left 'round,
    carrion for his dinner.
But it's me he's focused on, coaxing me into conversation.

When I respond, he flies closer, perches to stare at me.
Beady black eyes glow fiery against shiny indigo feathers.
He caws in clusters of three, his wings inflating with air
    upon each inhale,
Cccaaaawwww, cccaaawww, ccccaaawwww.
My responding calls intrigue him, and we converse.
An arc of connection cuts through the applecrisp autumn air.
He pauses, turning his head to an improbable angle
    to suck water through his long thin beak
    from the shallow pool
    puddled in the cement birdbath.

I wonder is he a bird of sorrow, or a creature of connection?
    I know he is both.

I learned this duality from my father, a man whose lionheart
    was far too big for the losses his love suffered in its youth.
He was a Jewish man who loaned his collection of Santa Clauses
    out to a different shopkeeper each year,
    so that his whole West Village neighborhood
    could enjoy the riotously diverse yet uniform
    red-and-whiteness of them.
His emotional voice was always tentative, feelings muffled
to lessen the risk he took in expressing them—
    his loss of love twisted into fearfulness of losing face.

But felt, his affection was a soothing bath that made everything safe,
    the sun I basked in when I was small.
Arms I could count on—except for those contracted moments
    when the shadow of loss overtook him.

Isn't it strange, how unspeakably beautiful life becomes
    whenever death draws near?

It hovers close now, all the time, with extinctions everywhere,
eighteen hundred species disappearing daily—
    my mind reels at it, staggering.
The tundra melting, trees tilting drunkenly as they lose their ground,
entire cultures losing their lifeways, the terrain too erratic
    for hunting anymore.

Who mourns these losses?

To enter that one-ness, the kinship of the crow,
    we must first feel the pain.

When I knew I'd never have a child, I vowed I'd have myself.
I wept with relief when the wise man told me
I had more children to care for in this life
    than I could if I had my own.

How can we close the circle, complete the cycle,
    and not go mad with grief?
Afraid I'll start wailing, I rock inwardly and don't stop.
My body moves with the sadness in waves that offer me comfort.
My rocking helps to still my maddening mind.

How can we grieve for the vividly colored corals bleached white,
    for the elephants brutally hunted for their tusks,
    for all those whose habitats have been logged
    to make mail-order catalogs, phonebooks, and newspapers?

The crone within me wants to shake us all awake, screeching:
    Don't you get it?
This is no time for small talk
This is a time for mythmaking
This is a time for epic poetry
This is a time to tell the tales of life, love, and resilience
    that'll become our compass for the days ahead.
A time to remember the grace and celebrate the magic
    that infuses and informs this world.

We live on the only planet we know of where the sun and moon
    appear the same size.
The only planet where an eclipse is possible.
Doesn't that seem like instructions to you?

To awaken from this self-induced slumber,
    to emerge from this contracted isolation,
    we've got to drink down the darkness
    and dive to our deepest fathoms.
Peel off our fancy garments of presumed protection,
    to land at the bottom, naked, cold, and bruised,
    with nowhere to go but up.

Time we shed the venom that got us here,
    the red rage of blame and shame.
And choose instead the anger that rises,
    pure and clean, up through our feet.

NINA SIMONS   221

That draws us to our full height, knowing what must be done,
    clear about what must be stopped,
        sparking us to stand for what we love.

How else can we begin the healing?
The web that holds our world together is tattered,
    with all our hopes and dreams suspended in it.
No sutures, butterfly closures, or Elmer's glue can fix it.
Only our tears can begin to mend its tattered strands,
    tears and giving ourselves to keening, pining, grieving.
Mourning how much is dying, mourning so that the light can return.

The revolution must have dancing; the women know this.
The music will light our hearts with fire,
    the stories will bathe our dreams in honey
    and fill our bellies with stars.
The interlacing of our souls will infuse and renew our humanity.
Our rhythms will merge with the heartbeat of the earth.

What breaks the mourning open for me?
It shines through my connections, my friends, my kin.
Some that are human, and some that are not.
I soar in the sea, glide stealthily among sea turtles,
    swoop over snowpack, eagle shushing.
I laze lizard-like on warm boulders amid frigid rivers,
    slurping oysters gathered fresh from warm, moist sand.

I am lifted by the courageous uprising of women, and girls,
    and of the emerging voice of the feminine within us all.
Together, leading from our feeling side first,
    we may yet restore balance before this precarious disc
    of our civilization
tumbles over the precipice.

And I am strengthened by my kinship with the land,
     with the high desert hills of Northern New Mexico.
Her mountains first called to me twenty-five years ago,
     and we've barely stopped talking since.
She reminds me of a time when her desert landscape
     was submerged underneath a shallow sea.
I visit her alluvial fan, a place where her rocky ridges
     meet a flattened plain.
A riverbed splays there, opening her legs to a widened basin.
A great open hand of sand is mounded there,
     to mark the fertile zone where two ecosystems meet.
Tickled by the magic of landing on this fulcrum,
this place where the differing worlds meet,
I pray for the help of the invisible hands
     of those who came before.

At dusk, I wander down the arroyo by our home.
I am flanked by criss-crossing dogs chasing scents,
     and a crow swoops low over my left shoulder, cawing.

At the bottom I stop, standing still on a sandy spit
     savoring the dry, clean scent of ponderosa forest.
Near my feet, a perfect white shell catches my eye.
It spirals pristinely, speaking to me in soft and sacred whispers.
Listening closely, I hear stories of its life before,
     and of its mother, the shallow sea.
I know the sea as my mother, too.
She holds me softly when I feel empty,
     and tickles me gently until I find laughter again.

My gratitude for the beauty of this world fills the spaciousness
     within me, and I begin to understand the crow's complexity,
and to see its embodiment of Kali, goddess of death and rebirth.

Within me, I feel the dualistic dichotomy
    of connection and disintegration melting away.
I remember how daybreak follows mourning.
They are waves in an ever-changing sea
    that together define the tide.

❧

Nina Simons is a social entrepreneur and co-founder of Bioneers, a non-profit that features breakthrough solutions for people and planet. She translates her life experiences into tools for serving the emerging leadership of others. Simons currently focuses largely on writing and teaching about women's leadership and restoring the feminine within us all—and on leveraging Bioneers' inspiring solutions and stories to transform how we live on Earth and with each other. See www.bioneers.org.

Waking Up
from Despair

OPEYEMI PARHAM

Being a retired family doctor, I use medical language in the story of
"How I Am Living, Now."

The culture into which I was born is dying, as we reap the consequences of our own greed, selfishness, and unsound ecological choices.
As I became aware of this, I reacted with the same psychological responses
that one has to one's own or a loved one's death: denial, anger, bargaining, depression, and acceptance.

Denial

I was born into a time of great promise, transition, and change. I am an
African American woman raised in an upper-middle-class-but-Bohemian-
leaning family. Having grown up in an integrated environment, I chose
my friends looking beyond skin color or religion, and I dated across color
lines. I went into an idealistic government volunteer program after college. I enjoyed shocking others with the complexity of who I was, and
by my refusal to fit into a box or behind a label.

Then I began to wake up to the poor choices that my country was
making and the consequences of those poor choices to the lives of individuals, and the future of human life, on this beautiful planet.

Like a woman who finds a lump in her breast, I felt it first as a growing apprehension. I had enrolled in medical school at Howard University, the oldest traditionally Black institution for higher learning in this
county. As a "twenty-something" marriage-age female, I found myself
isolated from my Black peers and confused by the lack of "quality" Black
men at my college. Where were my socially progressive and radically

thoughtful peers? All my medical student peers seemed interested in was how much money they could make. It was in my second year, when I did a special rotation at the public hospital across town, that I found all the brilliant, beautiful, politically active men that I expected to find at the University.

They were in the prison section.

The shock of this observation broke through my denial, and launched me into the next phase.

Anger

My anger was directed against "the system." I stepped up my personal political activities intended to subvert that rigid racially and class-segregated system from within. I married an Irish-Catholic man and had two children with him. While working as a physician, I made myself available to marginalized and disempowered groups as much as possible. My anger was a powerful engine, fueling my work and fueling it well.

Something else was creeping into my political consciousness: environmental awareness. I had read Rachel Carson's *Silent Spring,* that troubling and inspirational work by the marine biologist who advanced the global environmental movement with the book's publication in 1962. In *Silent Spring,* Ms. Carson asked questions about potential connections among the increased rates of physical deformities in marine creatures lower on the food chain, pesticides, human cancer risks, and potential ecological catastrophe. Her decision to write this book emerged from her own experience of breast cancer. As a physician, I could not ignore the connections I saw between environmental toxins and the breast cancers, brain tumors, and leukemia in my patients. As I cared for those living with and dying from environmental toxins, I began awakening to another unhealthy aspect of being American—our lack of deep spiritual belief and connection.

My definition of "environmental toxin" expanded from literal to figurative. Frenetic levels of stress contributed high blood pressure, and heart attacks. Socially supported addictions of workaholism, poor nutrition, and intense consumerism connected to the high rates of anxiety, obesity, and depression. I began to see how unhealthy the lifestyle of a "successful" American truly is.

I can look back now and see another emotion, festering beneath the anger, that I did not see at the time. That emotion was fear.

Bargaining

At age thirty-five I looked around me and saw that, despite my personal political attempts at change, life in America looked worse than ever. During this phase I would wake up to some new insight and bargain my way back to complacency. Did my country actually have an election (or two) *stolen*? Well, I would become a serious tax resister, and that would allow me to continue to speak my conscience, counterbalancing the nightmare of my country's growing fascism, increasing heartlessness toward developing-nation disasters of AIDS, famine, religious wars, and despotism, and the pervasive American denial of accumulating evidence on global warming.

While the social and political realities were frightening enough, my science background combined with the medical stories in my day-to-day practice forced me to pay attention to preventative health issues related to the environment. The wake-up call led by Helen Caldicott's group Physicians for Social Responsibility stressed the insanity of focusing on vaccinations in Africa while simultaneously ignoring the threat of a nuclear winter to the health of all humankind.

A second thread began to emerge in my life. On Earth Day in 1989, I danced the Spiral Dance with hundreds of Unitarian Universalist women. Starhawk led the dance, directing us to listen to the earth. "If you listen carefully," she said, "it just might speak to you." I had a profound spiritual awakening in that moment of Deep Listening.

From that moment forward, the health and well-being of Mother Earth—Gaia itself—became my social responsibility.

I embraced an earth-based spiritual ethic. As my ecological/environmental awareness blossomed, with that unnamed emotion of fear beginning to churn and work itself inside me, it was this spirituality that kept me sane. I struggled with increased concerns about the cost of my lifestyle—my ecological footprint—to others. I began to wake up to what types of "soul nourishment" actually fed me. I found that more and more solace in my life came to me in natural settings. I went to fire circles instead of malls and cineplexes. I grew to prefer being deep in the woods to being in city life. I stopped watching television and began to watch the night sky and the cycles of the moon and stars.

One fact was indisputable. *Something* was dying. Whether what was dying was the oppressive dominating cultural ethic, the earth, or humanity itself was not clear.

Depression/Despair

When despair hit me, it hit suddenly, fast and hard.

I was a physician in a culture of caretaking, with little to no support for the caretakers. I had no community of like-minded peers with whom to share my growing concerns and fears; I had become totally isolated. Five years post 9/11, in the spring of 2005 (post-tsunami, pre-hurricane Katrina), I felt overwhelming guilt at the gluttony of American consumers, and I felt deeply disturbed by our toxic effect on our environment. I lost faith that anything that I was doing as a physician was making any difference at all to the long term health and safety of my children and that of my children's children. I saw humanity headed for a full-blown, global ecological catastrophe, and I felt totally ineffective at making any change of substance. I did not want to witness or participate in whatever happened next. I felt so tired; I craved death, where I could be safe in the deep dark womb of Mother Earth.

I gave up, and I attempted suicide.

Acceptance

But miracles do happen, and I survived my own Dark Night of the Soul.

I did not simply live through my descent into my own terror. I came out the other side, healed in ways that are difficult to describe. I underwent a spiritual transformation.

As a consequence of my suicide attempt, I lost my professional power and position as a conventional MD. Ironically, that professional loss finally shook me loose from my last attachments to an outdated and overly consumptive lifestyle. Now I live very simply, with few possessions. A state of grace is always just a heartbeat away, as I recall and I appreciate the miracle of Being Alive. I feel it in the natural world around me, as I walk a daily circuit that takes me down country roads, past pastures with horses and houses with friendly dogs, over the rushing water of brooks, through wooded areas with birdsong in the air. I allow my day to unfold around me, rather than rushing out to meet and control it.

What I see is my own life, holographically represented in the Larger Scheme of Things. I lived the first half of my life as many Americans do—out of touch and out of balance with *my heart* and with Nature.

I sit, in the present moment, with priorities vastly shifted.

Life Today

I used to act for political change out of anger, with a sense of long-suffering sacrifice. Underneath that anger was a fear—a pernicious skepticism—that nothing would ever actually change. Now, I act from my heart, with renewed faith and a profound belief in miracles. I continue to make my personal life political. Post-"conventional doctoring," I am learning about herbal healing in a program that nourishes my intuitive skills while also honoring my intellect. I appreciate the importance of "knowing the herbs in my own backyard," literally and figuratively. I work in my local co-op as a clerk. My knowledge of dietary supplements, vitamins, and homeopathy grows as I witness customers waking up to Ecological Reality. I support their choices—simple ones, such as buying products marked with bright labeling "local heroes" (produced, raised, or grown right here in Western Massachusetts); these products used less gasoline and therefore emitted less carbon dioxide in their production. And there are more convoluted choices leading many of us to work together on building and sustaining community.

While I live now with less trust in my government, I find that I have more faith in my neighbors and my community. I am excited by our Five Rivers Council. This council (aspects inspired by Starhawk's book *The Fifth Sacred Thing*) is committed to creating a sustainable model of living for our small corner of New England, one that is reproducible in other communities. We educate our neighbors on a variety of issues: each person's impact on our local environment, emergency preparedness for New England's likely "eco-logical consequences" of global warming (floods, blizzards, and "microbursts"), helping local farmers stay viable and "go organic," awareness of biodiesel plusses and minuses, awareness of solar and wind possibilities for local energy sources, eating locally and seasonally, and working to shut down our nearest nuclear power plant (one of the oldest in the nation).

I appreciate what is here now. I count the blessings of a life being lived well.

What must die is the dominator culture. The one we have created, that is so very clearly unhealthy for us all. And it *is* dying. We humans can die along with it, or we can change. The Five Rivers Council in my county helps others wake up as I have awakened. While the hope is that the awakening can be less dramatic than mine was, maintaining the status quo is no longer an option.

My spirituality and faith in humanity (despite rationally deduced reasoning that suggests otherwise) sustains me. I release myself from a Judeo-Christian mythology that has eye-for-an-eye credos, a justification of suffering, and original sin at its center. I invite the possibility that a gentle Universe awaits humanity's transition from adolescence to adulthood, as we accept stewardship of our planet and empathic connection with one another.

My small "w" world as I knew it is dead. I look beyond my own story of personal catastrophe and see global warming, here and now. Some call it ecological catastrophe. My big "W" World as I know it is dying. A great transition is *here*. We humans will come into "full catastrophe living"—Jon Kabat-Zinn's phrase for the kind of moment-to-moment mindfulness that is necessary to a sane and centered life—from our hearts, or we will not.

How do we greet this great transition? That is what we are discovering, as we live, as we breathe, as we dance each moment.

I choose to feel power in the earth as it responds and reacts to humanity's actions. I choose to take my fear and breathe it into excitement. The earth, older than I can even imagine, is *reshaping itself.* The earth is not dying; the earth will survive. I am optimistic and hopeful that human beings will survive, too.

❧

Opeyemi Parham is now a "feral" (once domesticated, now returned to the wild) physician. She works as a clerk in a local health foods store, in her home community of Greenfield, in Western Massachusetts. Opeyemi continues to pursue her vocation of medicine woman, using natural herbs, health education, and ceremony, having evolved beyond the constraints of her allopathic training. In her spare time, she sings, dances, and Speaks Truth to Power through her Web site (www.ceremonyheals.com) and her radio show, "Channeling the Muse."

River Gods

KEN LAMBERTON

For me, living in the Southwest, it's a question of finding a way. The poet and writer Alison Deming says, "In the desert, one finds the way by tracing the aftermath of water," so I look to the dry rivers. And there I find people working to restore them.

On the afternoon of the winter solstice, Daniel Preston and Renee Red Dog of the Tohono O'odham Nation stand in the rain next to the Santa Cruz River thirty miles north of Tucson. The smell of burning sage mingles with the tonic scent of wet creosote. Fifteen observers, jacketed and sweatshirted against the chill, huddle around a potted blue paloverde sapling. As Daniel blesses the tree in the O'odham language, a volunteer slips it into the ground.

Daniel, wearing his trademark bolo, speaks about how his ancestors once drew life from the Santa Cruz and how life has begun to return here now, coming full circle. He asks everyone to face east and then he and Renee begin to sing, praying for strength and guidance for the people who are working diligently to heal the land.

The Santa Cruz, river of the Holy Cross, is a dead river, a dry channel of shining quartz grit except during heavy storms. It's been this way for more than a hundred years, since groundwater pumping for agriculture and mining lowered the area's aquifer. Occasional floodwaters subsequently downcut and entrenched the river's once-meandering course.

But here, downstream from Arizona's second-largest city at a place called the Simpson Site, the river flows, unfurling a dark liquid ribbon along a cottonwood- and willow-hemmed seam. Even during the hottest of summers, the air smells of dust and effluent. Tucson's unwanted wastewater has granted the Santa Cruz a second chance.

Standing with Daniel Preston and Renee Red Dog is Kendall Kroe-sen, Tucson Audubon Society's Restoration Program Manager and the tall, lanky, bearded force behind the salvation of the river. Kendall is a permaculture specialist whose handiwork appears at many sites along the river in mesquite-clotted basins and swales. His interests in people and nature, and his research into what makes communities successful, have led him to seek ways to create a more sustainable human society, particularly in the Sonoran Desert. Kendall sees ominous trends in today's commerce-driven society, which consumes nonrenewable resources along with cultural diversity. His passion is the belief that natural ecosystems and wildlife can thrive alongside humanity.

Rodd Lancaster is Kendall's disciple and right arm, a quiet, taciturn man who cultivates vibrant creosote bushes and paloverde trees out of dry hardpan. Like his plants, he lives "off-main." His home and amazing garden are supplied by harvested rainwater rather than city pipes. This, in a desert that receives less than twelve inches of precipitation each year.

I like to think that I might live my life like Rodd and Kendall, tread-ing lightly on the earth while engaging with people and nature to help create healthy communities. It's not easy. The least I can do is to par-ticipate in their stories. Theirs are stories of hope. Stories of despair and defeat have already been told, and we have enough of them. Despair is why people allow our rivers to be drained and their beds used as land-fills. But hope, unlike despair, leads to productive action. Hope means another world might be possible.

At the Rio Cocóspera in northern Mexico, enormous cottonwoods rise out of the riverbanks and erupt into the sky like glaucous-plumed thun-derheads. Male vermilion flycatchers pump their dark and undersized wings to corkscrew slowly above the highest branches. Everywhere is the smell of water and rock, a quiet gathering of pungency, of shining run-nels and algal slackwater trapped in the wallows and trackways of cattle. I look north toward the Sierra de Pintos, whose upwelling flanks bend the river south toward the distant Sierra Azul. Here, researchers recently captured a wild ocelot on film, collected from a camera that my oldest daughter Jessica helped set up. Trotting along a rocky drainage only twenty-five miles from Arizona's border with Mexico, the endangered spotted cat was a pleasant surprise. The last ocelot was documented in the region in 1964. Jessica says the animals, like their jaguar cousins, may be crossing from Mexico into Arizona, perhaps even using the Rio Cocóspera as a pathway north.

My daughter, in addition to her studies at the University of Arizona, is working with the Sky Island Alliance, assisting international efforts to record the movements of our border-crossing jaguars. The nonprofit conservation organization teams with scientists, volunteers, and landowners to protect and restore the region's diverse native species and habitat. Jessica has been traveling into remote country in Arizona and Mexico with Sky Island staff to set up motion-sensor cameras, hiking and camping with men and women who spend most of their days away from modern conveniences like showers and toilets. It was Sky Island Alliance that brought Jessica and me to this beautiful and wild borderland country of desert rivers.

Like the nearby seventeenth-century Kino mission by the same name, the Cocóspera has lost much of its former glory. At first glance the river seems to be only a shallow irrigation ditch in the service of livestock and agriculture. But soon we see more than cows and alfalfa fields. A cottonwood grove stretches down the far bank, leaving the near side open to sky. A gray hawk screams among the branches. In its dark robe of feathers, a great blue heron slowly toes the algae-matted river margin, the bird's huge feet leaving Triassic prints in the mud.

Unmistakable signs show the river's heavy use. Cattle have trampled its banks. Canals divert its water. But, unlike many rivers in the Southwest—the Gila, San Pedro, and Santa Cruz rivers, for example—water still flows year-round in the Rio Cocóspera. Erosion hasn't downcut the river and trapped it into a steep-walled channel reinforced with gunnite. Native fish, like the longfin dace, continue to send ripples across its glassy surface. Rivers will probably always be in the service of people. The question, for Mexicans and Americans alike, is will that service be sustainable?

One Mexican rancher is already answering that question. Carlos Robles owns Rancho El Aribabi, a ten-thousand-acre spread that reaches
from the Río Cocóspera into the Sierra Azul mountains. His ranch house
rests on a saguaro-stabbed hillside above the river, which today is a
fencerow of cottonwoods and willows that screens the corrugated eastern horizon. Ten years ago there were no trees, and the river was a stink-
ing, muddy ditch of eroded banks and a dead wickerwork of Bermuda
grass. Then Carlos removed his cows.

Jessica and I arrive at El Aribabi to join a group of Sky Island Alliance
(SIA) researchers and volunteers who are conducting a wildlife inven-
tory of the ranch. Our invitation had come from Sergio Avila, an SIA
biologist who was born in Mexico City and grew up in Zacatecas at the
southern end of the Chihuahuan Desert. With a master's degree from
the Universidad de Baja California, Sergio has worked with wildlife rang-
ing from mountain lions to sea lions.

We spend our first day hiking rugged canyons, searching for mammal tracks and changing the film in several motion-sensitive cameras.
Sergio leads the way, while Cynthia Wolf, a freelance biologist, wilderness outfitter, and tracker extraordinaire, scans the sandy soil and mud
for animal sign. Jessica keeps notes for each camera and monitors our
GPS coordinates.

In the evening, we rejoin Carlos at his ranch house and descend upon
his kitchen. While Sergio reheats some *carne asada* he finds in the refrigerator, David and Greg, two volunteers from Nogales, Arizona, prepare
what they call *"rajas de chile verde en crema,"* sautéing onions with roasted
chilies, adding cheese and a can of Media Crema. We eat the food, standing around Carlos' gas stove, warming ourselves next to the grilling
tortillas. After dinner, Carlos brings out a bottle of El Jimador, pours
shots, and passes them around. Then, raising his glass, he says in a loud
voice, *"Al futuro del jaguar y el ocelote,"* offering a toast to the future of
two amazing animals only recently seen on his ranch.

The next morning, Carlos talks about his business, how he has moved
away from cattle ranching and begun to diversify into less environmentally destructive activities such as hunting and ecotourism. He has turned

his ranch house into a retreat center, its many bedrooms with fireplaces, its central kiva-like living room and open porches refitted for large or small groups of students, researchers, and sportsmen. "I can do anything people want," he tells me.

"Horseback riding with a barbecue?" I ask.

"*Sí*. And campouts, nature and bird tours, guided hunts."

"What about a bed and breakfast so I can bring my wife?"

"Sí."

El Aribabi is one of the premier ranches in northern Mexico, hosting at least thirty endangered and/or threatened species. The ranch boasts 165 species of birds, including elegant trogons and rose-throated becards. Green rat snakes and Gila monsters patrol the riverbanks where pig-like javelinas come to drink. Troops of coati, a long-snouted mammal that looks like a cross between a raccoon and a spider monkey, root in the undergrowth for grubs and tubers. And there are trophy-class Coue's white-tailed deer, whose antlered bucks attract hunters from all over the world.

Recently, Carlos Robles and Sky Island Alliance entered a groundbreaking international collaboration with a Memorandum of Understanding. "It's more than an agreement not to shoot predators," Sergio explains to me on our last day. "It outlines our shared goals about conservation and will help us shape strategies to restore and protect the rivers and desert of El Aribabi."

Jessica, who participated in the signing celebration at El Aribabi, says the event marks a new and hopeful outlook on conservation across international borders. "It is so inspiring to see people working together to make the world a more diverse and beautiful place," she said, "especially people divided by so many barriers: languages, lifestyles, and fences."

It was the path of water that originally drew Native peoples to the Southwest, water for life and water for direction in life. It was the same for all who would follow them, whether Spanish or Mexican or American. Water creates community. And sometimes more. Writer Peggy Schumaker says that water in the desert is always holy. If this is true, then people like

Rodd and Kendall, Carlos and Sergio are holy men, the first true holy men of rivers I've met.

❧

When Ken Lamberton published his first book, *Wilderness and Razor Wire* (Mercury House, 2000), the *San Francisco Chronicle* called it "entirely original: an edgy, ferocious, subtly complex collection of essays." The book won the 2002 John Burroughs Medal for outstanding nature writing. He has published four books and more than a hundred articles and essays in publications such as the *Los Angeles Times, Arizona Highways,* the *Gettysburg Review,* and *The Best American Science and Nature Writing 2000.* In 2007, he won a Soros Justice Fellowship for his fourth book, *Time of Grace: Thoughts on Nature, Family, and the Politics of Crime and Punishment* (University of Arizona Press, 2007). He has just completed a book about southern Arizona's "Dry River," the Santa Cruz. He holds degrees in biology and creative writing from the University of Arizona and lives with his wife in an 1890s stone cottage near Bisbee. His Web site is www.kenlamberton.com.

Questions for
a Sacred Life

BODHI BE

It's a good day to die!

We as humans have two very important things in common.
Our bodies will die, and, we don't know when.
At least 90 percent of the people who will die today
did not know yesterday that they only had one more day to live.
Take a moment and feel that.

Can any one of us guarantee we'll be alive tomorrow?
Could this be at the core of our individual and cultural denial,
our avoidance, leading us to consume and destroy the world, . . .
to be anywhere but here in this present moment where we might
    make contact
with our feelings of uncertainty?

Crazy Horse reportedly spoke the words "It's a good day to die!,"
to his assembled warriors as they were about to enter into battle to
    protect
what was left of their families and tribes.

What does it mean to be able to say "It's a good day to die"?
Are any one of us ready to die today?
What would a "good day to die" look like?
Would you bring everything to the table, your "A game" if you
    knew
today was the day?

How much of our attention would we give to each moment of the
    day?
Overcoming our cultural conditioning, would this very moment be
    enough, as is?
Will we finally be enough, as we are?
What would you choose to do in your life, in your relationships, in
    your world,
for it to be a "good day to die"?
What "unfinished business" might you address?

The dominant culture in the world today, transcending
    nationalities and borders, is a culture out of balance, which has
    lead to a people out of balance.
A people out of balance with our place in the web of all life,
    thinking we are, somehow,
above and exempt from the laws of the natural world.
Out of balance with our soul's purpose and the voice of God
    speaking within us,
We are out of touch with who we truly are and why we're here.

What is the connection between our fear, aversion, denial, and
    avoidance of death
and the condition of our world and our lives?
What happens to a people who view death as a failure, a mistake?
Something we rarely speak about, certainly not in front of the kids,
surely not in party conversation. "Oh, don't be morbid!"
How do these attitudes affect the ways we treat old people,
people who are dying, the dead?
How do these attitudes affect the way we treat the earth and its
    creatures?

When we have an "out of balance" relationship with death, we
    cannot help but be a people
out of balance with life and the life of the earth.

We hear about people dying everyday all over the world,
but how different it feels when someone close to us dies.
There's a crack in our world, everything gets very fragile.
Funerals and memorial services can be profoundly bonding experi-
    ences for a community.
Everyone gets very real when death is close.

What would life be like when the next breath is a precious gift?
Would we have more appreciation, more gratitude for our lives?
Living with the truth of not knowing how much time we have in
    this life,
would we treat our friends and families, our neighbors, the earth
    itself,
with more care and kindness?
Would our "stuff" hold such importance?

How do we, as a culture, currently deal with death when it does
    make an appearance?

When someone dies, our tendency is to immediately call someone
    to remove the body.
We are uncomfortable around death.
We'd rather ask strangers to take the body away so the next time
    we see it, it is embalmed
and made to look life-like, or we see an urn filled with ashes.
Only in the last hundred years have we given death care over to a
    "funeral industry"
and in doing so we have robbed ourselves and our communities,
    having removed ourselves
from participating in this aspect of life and its teachings.
What part does this play in a disposable culture that worships
    youth and anti-aging, and has relegated
our elders, as wisdom keepers, to the "elderly"?

There is a movement afoot, a natural death care movement intent
    on families and communities
reclaiming the work of caring for our dead, not paving a new trail,
    but simply clearing the weeds from a trail that's been around
    since the beginning.
Just as many of us helped to bring the birthing of our children
    back into our homes, so too, now "spiritual midwives to the
    dying" are assisting the dying and their families.

Recognizing this, not only as ours to do, as a community, we also
    recognize it as "sacred work."
We stand in the truth that who we are doesn't die, our bodies die.
Our willingness to be completely present and open offers an
    opportunity for healing and growth,
and often a way to support a soul moving on.
Being around dying people and dead bodies is powerful classroom
    for putting our lives in perspective.

What does natural, spiritual death care look like?

Once someone dies we may bathe them, oil and scent them, brush
their hair, and clothe them. We can transform their bed into an
altar, with flowers strewn about the body. We may invite friends
and family to sing, pray, laugh, cry and tell stories in the presence
of the body, and perhaps, the hovering spirit. Maybe we'd just sit
in silence. We may pick up a cardboard casket from the mortuary,
decorate and write prayers and mantras, on its sides *(shipping
instructions!)*. We lift the body into the prepared casket, then load
it into our van, all covered in flowers and caravan to the mortuary.
When it is time for the cremation, we lift the body into the oven,
and praying for good passage, start the fire. The time is coming
when it will be common for us to dig the hole for the body and
plant a tree over it.

We embrace the opportunity to honor someone's soul, their life, and this passage, by caring for and honoring the body.
It's powerful, it's intense, it's deep.
It's also quite ordinary.
It's what humans have been involved in for thousands of years.
In reclaiming this part of life, Life itself becomes more real and precious.
Having a healthy relationship towards death helps us be more alive.
It helps us to honor all that is alive.
Some may wonder "Is there life after death?."
Sometimes the more important question is, "Was there life before death?"
It's a good day to be alive!

Bodhi Be is an Interfaith/Innerfaith Minister, Hospice Volunteer, Funeral Director and Executive Director of Doorway into Light, a non-profit in the field of death and dying. He is the host of a weekly radio program in Hawaii called "Conversations with Death." Bodhi is a senior teacher in the Sufi Ruhaniat International, the Sufi lineage of Inayat Khan and Samuel Lewis.

He lives with his wife of twenty-five years on Maui where they homestead an organic garden and orchard, watch the sun make electricity, and play with their first grandchild. His Web site is www.doorwayintolight.org.

To Do the Will of God, Come What May

ALICE WALKER

This is what Martin said is important and necessary and what he would do.

I woke this morning feeling the same.

Though my "God" is Everything. Without boundary. Everywhere.

I looked out and a wave was crashing over the reef in front of my house: God. I looked up at the old gnarled tree just by the hedge: God. The hedge itself. Myself so small in the great God vastness as to be almost not here. Not present and yet I am here, present. Conscious.

It is a great gift to be a part of Godness.

It is Love—Godness itself—that gave me this vantage point.

April 10, 2008

CenterHina

Molokai

To What Purpose?

Only to admire, to praise, as I so often feel?

To marvel.

Wonders are endless and though there is suffering—so frequently human caused—there seems little reason to ever complain.

How did this happen? What mystery not to ever know.

So like butterflies we are, ever on our way to lightness, to flight, but without clue to destination.

I find this suits me.

That I am aware of floating through time and space gifted with the bounties of the journey yet not ever owning any of them. It is like living in a dream where everything seems real, solid, and yet we are, all of us—leaves, toads, humans—just passing through.

The Universe, the Cosmos, so vast. Time so vast. Surely we are recycled millions of incarnations as everything there is. Freedom to Perfect!

Seen from this perspective our suffering on this small planet is about learning enjoyment. Choosing peace over pain and destruction. Growing into a comfortable universality. Letting go of pettiness. Dissolving tribalism, nationality, speciesism.

I knew this as a child. That the daffodil might be me. Moondust. Barnacles in the sea. Rocks and bear claws.

Isis knew this. Humanity after Her forced not to know. Humans choose Gods small enough to wear like amulets to assuage their fear. But the wise do not choose a God because to them God/Goddess is seamless and is already wearing you.

Back from *that* Paradise—CenterHina, Molokai. All was as perfect as Life gets for us humans. Heat, coolness, mangoes, friendship. The last weekend we hiked to Halawa Falls and I went into the pool at the base of the Falls and swam toward Her, towering, rushing, spraying, over me. To be a part of all this! Sometimes my gratitude is almost more than I can bear. I bear it, often, weeping. As now.

❧

Alice Walker is an American novelist, short-story writer, poet, essayist, and activist. Her most famous novel, *The Color Purple,* was awarded the Pulitzer Prize and the National Book Award in 1983. Alice Walker's creative vision is rooted in the economic hardship, racial terrorism, and folk wisdom of African American life and culture, particularly in the rural South. Her writing explores multidimensional kinships among women, among men and women, and among humans and animals, and embraces the redemptive power of social, spiritual, and political revolution.

CHAPTER EIGHT

Hope in
Challenging Times

To Endure Climate Chaos, Live Dangerously and Cultivate Hope

BRIAN TOKAR

There is enough uncertainty in our lives today to engender a sense of profound unease in even the steadiest of minds. Vast upheavals in climate, economy, and society are upon us. The once moderately predictable patterns of weather, the seasons, heat and cold, moisture and dryness are falling out of balance. If we're paying close attention, we read almost constantly of heightened natural disasters and people uprooted from their homes and livelihoods. We ponder the latest predictions of climate scientists and often find ourselves aghast at the magnitude of the earthly changes that are likely to come.

While much of the world is already experiencing some of the severely destabilizing consequences of global climate disruption,* our experience in the northern tier of the United States and southern Canada is considerably different. Here, we find ourselves in one of the few places on earth where the near- to medium-term consequences of global warming may actually appear positive. The weather is often more extreme, and storms are less frequent but more intense; still overall warming temperatures appear to bring us some immediate benefits. Where I live, in the northern hills of Vermont, our growing season has lengthened from just over three months to well over five in recent years. Perhaps this accident

---

* See, for example, World Resources Institute, *Synthesis: Ecosystems and Human Well-Being,* A Report of the Millennium Ecosystem Assessment, Washington, DC: Island Press, 2005; Intergovernmental Panel on Climate Change, Working Group II Report, "Impacts, Adaptation and Vulnerability," 2007, available from http://www.ipcc.ch; *Human Development Report 2007/2008: Fighting Climate Change: Human Solidarity in a Divided World,* United Nations Development Program, 2007.

of circumstance will help us renew our connection to the land, and also encourage others to find ways to do the same. Perhaps this improbable opportunity compels us to help illuminate a journey away from impending catastrophe and, just possibly, toward a more harmonious relationship with the rest of life on this fragile earth.

Here in Vermont, the ethic of *simple living* has a rich and illustrious history. Living close to the land has long been associated with a high quality of life. Many of my friends and neighbors have an intuitive understanding that living more sustainably doesn't mean buying lots of fashionable "green" products or having all the latest high-tech "green" gadgets. Rather it means meeting more of our needs at home, sharing with our neighbors, and raising children who know the land even better than we do. It means homegrown tomatoes in the summer, myriad varieties of squash and potatoes, and all manner of home-frozen goods in the winter and, if we're meat eaters, knowing who raised the cows or lambs or chickens that we depend on for a rich but modest share of our diet.

Are small changes in personal lifeways sufficient to change the world and prevent catastrophe? Of course not, but small changes can add up to wider community- and even society-wide changes if they're carried out with clarity, intention, and a commitment to realizing a larger whole that's considerably more than the sum of the parts. Moving forward, from changes at the personal level to the community level and beyond, challenges us to reach beyond conventional expectations and create living examples of a richer quality of life that's considerably lower in material consumption. Eventually, we hope we can foster the kind of political reawakening that allows us to realize the much broader social and economic changes that are necessary. If we cannot, the consequences may be dire, as even the most environmentally conscious Americans contribute to a society that consumes many times its share, and often terrorizes the rest of the world with its hubris and arrogance.

So how do changes in our thinking come to inspire and activate changes in our lives and in the wider world? One fascinating example lies in the emerging local food and Slow Food movements. Slow Food founder Carlo Petrini has described how a contamination crisis in the

HOPE BENEATH OUR FEET

Italian wine industry created a need to change how local wines were marketed, and eventually led to a much broader transformation in the relationship between rural food producers and affluent consumers. For the first time, the food producers moved to the center of attention, "to make up for the low esteem they have hitherto enjoyed, and rewarded for their work in rescuing a species of livestock, a fruit or vegetable, a variety of cured meat or cheese."* Slow Food and its "locavore" counterpart in the United States have inspired many to experience the joys, as well as the challenges, of eating closer to home and knowing precisely where our food comes from. But neither movement can survive if it is reduced to just another elite fad, along with high-priced "green" products and "green" fashions. Local food advocates need to fully realize the potential for changes in personal taste to also systematically transform the underlying social relationships. We can no longer allow ourselves to be cast as mere "consumers" of the earth's bounties, nor continue permitting large commercial interests mediate between our communities and the people who grow our food.

Participation in growing our own food is also an important step. Gardening is already the most popular pastime of Americans, but the evolution beyond gardening for aesthetics, to a fuller and richer engagement in feeding ourselves, has only begun. Growing one's own food is, in Michael Pollan's words, a way to "heal the split between what you think and what you do, to commingle identities as consumer and producer and citizen." So here in Vermont, people are building modest greenhouses, transforming lawns into food gardens, and returning hillsides that were mowed for mainly aesthetic reasons back into productive grazing land, where animals eat grass, feed the soil, and ultimately become part of the solution rather than part of the problem. We have the highest proportion in the country of food that is purchased directly from farmers. At the same time, farmers' markets are also doubling and tripling in numbers in many urban areas, a profoundly hopeful sign.

---

* Carlo Petrini, *Slow Food: The Case for Taste,* New York: Columbia University Press, 2003, p. 51.

Similarly, some of my friends in the building trades are creating a whole new approach to natural building. While "green building" today often means massive, institutional buildings with sophisticated temperature control systems and all manner of newfangled synthetic materials, Vermont's new generation of natural builders is working almost entirely with wood, straw, clay and mud. They are experimenting with innovative truss systems and even mixing their own chemical-free paints. They'd much rather work with whole logs, with all their imperfections, than with factory-cut dimensional lumber. These are important steps beyond the world of mass production and consumption that has put humanity onto such a collision course with the cycles of nature.

My community is far from perfect. In Vermont we still have neighbors who live in a parallel universe of McMansions and SUVs, or struggle every day to make ends meet in leaky old houses, or in trailer parks. The cost of living is becoming prohibitive for many, and speculators continue to drive land prices into the stratosphere. But the urge to create a better way of life is widespread, and affects nearly everyone in important and inspiring ways.

Some forty-five years ago, as ecological science was just coming into its own, many thoughtful writers began to wonder whether there was something inherently subversive about thinking ecologically. Murray Bookchin, the founding philosopher of social ecology, responded with a resounding "Yes," arguing back in the mid-1960s that there is indeed something fundamentally revolutionary and transformative about an ecological understanding of the world. One of my own most important and inspiring teachers, he became a pioneering advocate for sustainable technologies, decentralized cities with politically empowered neighborhoods, a moral economy—freed from the built-in constraints and inequities of the competitive market—and, ultimately, a thorough recasting of modern humanity's relationship with the rest of nature.* Many of us who

---

* For an overview of the works of Murray Bookchin, see Janet Biehl, ed., *The Murray Bookchin Reader* (Montreal: Black Rose Books, 1999); also his *The Limits of the City* (Revised edition, Montreal: Black Rose Books, 1986), *The Ecology of*

were environmental and social activists gained considerable inspiration from this outlook, especially as movements arose from the grassroots during the late 1970s and early 1980s to end the first wave of nuclear power development in the United States. We organized ourselves into small, local "affinity groups" to develop regional antinuclear organizations from the bottom up, and also explored how an energy system that relies on the sun and wind could become the underpinning for a radically decentralized and directly democratic society. Those experiments in merging environmental opposition with a strong, reconstructive vision of a new society offer essential lessons for us today.

Today, it is clearer than ever that the obstacles to a free and sustainable society are social and political, not mainly technological. New ways to save energy and replace fossil fuels are invented and announced almost daily, and the means readily exist for virtually everyone to live a rich and satisfying life. Our established political institutions, however, continue to postpone these needed changes, and the portent of a future of deprivation and scarcity often looms large, defying all our well-fed hopes for a different outlook. Then along comes one of those exceptionally bright days that simply defy all the doom and gloom. They come most often in the Spring, but can arrive in any season. On those days, it is not merely necessary to act on the belief that we can help create a different kind of world—it actually appears to be within our grasp.

❧

Brian Tokar is an activist and author, director of the Vermont-based Institute for Social Ecology, and a lecturer in Environmental Studies at the University of Vermont. He is the author of *The Green Alternative* and *Earth for Sale;* edited two books on the politics of biotechnology, *Redesigning Life?* and *Gene Traders;* and co-edited the forthcoming collection, *Crisis in Food and Agriculture: Conflict, Resistance and Renewal* (Monthly

Freedom (Palo Alto, CA: Cheshire Books, 1982, currently available from AK Press in San Francisco), and "Market Economy or Moral Economy," in Bookchin, *The Modern Crisis* (Philadelphia: New Society Publishers, 1986).

Review Press). Tokar has been acclaimed as a leading critical voice for ecological activism since the 1980s and lectures widely on environmental issues and popular movements.

Fighting Fatalism
about War

JOHN HORGAN

As a youth with nothing to lose I indulged in fantasies about the end of the world. But I'm a father and a teacher now, and so I feel a moral obligation to be optimistic about humanity's future and to persuade others to be optimistic, too.

My focus lately has been on warfare, whereby we destroy not only each other but also nature, in ways both direct and indirect. As the Vietnam-era poster said, "War is bad for children and other living things." Fatalism about warfare is rampant now. I recently asked 205 students at the engineering school where I work the following question: "Will humans ever stop fighting wars, once and for all?" More than 90 percent answered "No." That is roughly the same response rate I've gotten from journalists, scientists, friends, neighbors, hockey teammates, cab drivers, and others I've polled over the past few years.

Justifying their negative responses, most people offer variations on Robert McNamara's remarks in the documentary *The Fog of War:* "I'm not so naïve or simplistic to believe that we can eliminate war," the former secretary of defense says. "We're not going to change human nature any time soon."

In my writing, public talks and private conversations, I've been trying to persuade people that fatalism about war is wrong on both empirical and moral grounds. Empirical because the historical and even pre-historical record shows that war—far from being an inevitable manifestation of our innate aggression—is a response to certain environmental and cultural conditions. Moral because the belief that war will never end helps to perpetuate it.

Recent scholarship on warfare seems, superficially, to support the view that war is inevitable. Just a few decades ago, many scholars believed in the myth of the peaceful savage, which depicts war as a by-product of modern civilization that did not exist in pre-state societies. Actually, recent research in archaeology and anthropology reveals that the vast majority of primitive, pre-state societies engaged in at least occasional warfare. Mortality rates from violence in some societies reached as high as 50 percent.

But these grim statistics yield a surprisingly upbeat message: things are getting better! Hard as it may be to believe, humanity has become much less violent than it used to be. In fact civilization, far from creating the problem of warfare, is apparently helping us to solve it. In the blood-soaked twentieth century, one hundred million men, women, and children died from war-related causes, including disease and famine. The total would have been two billion if our rates of violence had been as high as in the average primitive society.

Steven Pinker, a psychologist at Harvard, argues in a recent essay that "today we are probably living in the most peaceful moment of our species' time on earth." Conventional wars between the armies of two or more nations and even civil wars have decreased sharply over the past half-century, he points out, as have casualties. We are now dealing primarily with guerrilla wars, insurgencies, terrorism—or what political scientist John Mueller calls "the remnants of war." Noting that democracies rarely if ever wage war against each other, Mueller attributes the decline of warfare over the past fifty years at least in part to a surge in the number of democracies around the world—from twenty to almost one hundred.

These statistics do not provide much solace to the victims of violence in Iraq, Darfur, Sri Lanka, Palestine, Colombia, and other troubled regions, but they show that we are moving in the right direction. Other recent events offer more grounds for optimism. As recently as the late 1980s, we faced the threat of a global nuclear holocaust. Then, incredibly, the Soviet Union dissolved and the cold war ended peacefully. South Africa's apartheid also ended without significant violence, and human rights have advanced elsewhere around the world.

The Harvard biologist Edward O. Wilson, the world's leading champion of biodiversity, is confident we will find ways to cease making war on nature as well as on each other. "I'm optimistic about saving biodiversity," he told me recently. "And I think that once we face the problems underlying the origins of tribalism and religious extremism—face them frankly and look for the roots—then we'll find a solution to those too in terms of an informed international negotiation system."

The single most important way to advance the causes of conservation and peace, Wilson and other scientists say, is to improve the education of females in the developing world. Many studies have demonstrated that as female education increases, birth rates fall. A stabilized population decreases demands on governmental and medical services and depletion of natural resources and hence the likelihood of social unrest. A lower birth rate also reduces what some demographers call "bare branches," unmarried, unemployed young men, who are associated with higher rates of violent conflict both within and between nations.

I am still constantly trying to persuade people to view war as eradicable. I urge you to do the same. All living things will benefit in countless ways once we humans stop diverting so much of our energy and resources into fighting and preparing for wars. And the first step toward ending war and our assault on biodiversity is to believe that we can end it.

≈

John Horgan is a science journalist and Director of the Center for Science Writings at the Stevens Institute of Technology, Hoboken, New Jersey. A former senior writer at *Scientific American* (1986–1997), he has also written for the *New York Times, Time, Newsweek,* the *Washington Post,* the *Los Angeles Times, The New Republic, Slate, Discover,* the *London Times,* the *Times Literary Supplement, New Scientist,* and other publications around the world. Horgan is currently doing research on the widespread belief that human warfare is inevitable. His Web site is www.johnhorgan.org.

Little by Little

Every day as I skim the paper, my heart breaks with the endless news of escalating wars, disease, hunger, and a planet on the verge of environmental catastrophe. It's all very overwhelming and depressing. The headlines can easily create a feeling of hopelessness, but each morning, as I set them aside and move to my office to begin working, I turn to my favorite Haitian saying:

"Piti piti n a rive."

This is a Creole phrase that means, "little by little we will arrive." Remembering these words of hope refuels my spirit and reminds me that every small step we take towards change makes a difference.

I heard this saying for the first time on a muggy July afternoon in Port-au-Prince. I'd come to visit a food program for children that I'd helped start a few months earlier. We only had enough money for one meal per week, so food was served on Sunday afternoons. Five hundred children gathered to eat on the Sunday I was there. Some walked for miles. For most of them, it was their only hot meal of the week.

The meal was served by members of St. Clare's Church, who spent two days preparing huge pots of rice, beans, chicken, and a stew made with vegetables from the farmers' market. For hours, plate after plate was passed down a line of volunteers and placed in front of children who waited patiently at the tables. As they spooned in the meal and scraped the bottom of their bowls for every grain of rice, I watched their focused faces—chewing and swallowing with urgency and excitement. Clearly famished. Their eyes were serious, but many of them still had a sparkle. I watched the littlest ones keep up with their older siblings, completing gigantic portions in record time. Later, as they filed out of the St.

Clare's rectory, they looked happy and full, but I wondered what the rest of the week held for them and how long it would be before they all felt hungry again. I wondered if this tiny food program was really making any difference at all.

The next day, the priest of St. Clare's Church, Father Gerry, took me back to the food program site. He said he had something to show me. We walked across the dusty yard and stood next to the empty kitchen. The building, which had been packed with cooks and children the day before, was quiet. It was rarely used during the week, although Father Gerry announced he had big plans for it.

"This is where the outdoor cafeteria will be, a large space for the food program to be served. Hundreds of children will be able to sit down at once." His eyes smiled and his voice was convincing.

"An outdoor cafeteria?" I said confused. He nodded. "With a roof and sturdy tables and benches for the children and a concrete floor so they don't hurt themselves walking on the rocks and glass." He led me around the corner and pointed to the back of the building. "Right here is where the new kitchen will go. With running water and a big stove to cook food so we can serve meals to the children during the week."

I squinted in the sunlight, trying to imagine the new kitchen and the possibility of more days of food. He continued with a big smile. "Over here is where the school will go." He pointed to the empty half-acre lot to the left of the rectory. "With a daily lunch, and a library and health clinic."

"And over there," he spun around and pointed to the road leading to the rectory, "I see the roads paved. No more roads that wash away every time it rains. No more struggling up the hill." Then he turned slowly in a circle, pointing to the homes surrounding us. "Margaret, I see all the children fed and their parents working. Everyone has enough food to eat and electricity and running water."

I looked with him into the neighborhood, past the piles of garbage and the dark interiors of the dilapidated homes trying to see the vision he saw. But I couldn't. The bleak reality of the neighborhood was overwhelming. So I shut my eyes. Standing in the middle of the empty lot, I tilted my head back. The sun burned my cheeks as I tried to imagine a school next to the rectory. After a few seconds, it began to take shape in my mind's eye. It was three stories high with bright blue, orange, and yellow paint. Happy colors. I imagined a bell ringing and dozens of children skipping through the gate with books in their arms. They chatted and laughed as they went to their classrooms and sat behind new desks. Teachers greeted them and lessons began.

We stood in silence for a minute and then he continued, "We have a Creole saying I want to teach you. 'Piti piti n a rive'. That means little by little we will arrive. One step at a time, Margaret. In Haiti, sometimes they are very, very small steps. Sometimes they go backwards. But it's important to keep taking steps, even though they are small. Never give up. Never lose hope. One day, maybe not during my lifetime, but one day, we will get there."

Over the last seven years as our food program has grown, that July afternoon with Father Gerry continues to remind me of how important it is to have a big, inspiring vision and then to break it down into small steps. Sometimes I can get lost in the vision and forget that to get there requires action and a lot of patience. Other times I can get stuck in the action and forget the vision that's needed to inspire me to keep going. A healthy balance of the two is the key for me. So far, my work with Father Gerry and the What If? Foundation has not resulted in the vision he described—but we're headed in that direction—piti piti. We're up to five meals per week for nearly one thousand children. We haven't built that school, but the foundation pays the tuition for one hundred children to attend other neighborhood schools.

I remember my high school physics teacher saying in class one day that when a fly lands on a steel beam, the beam bends. It's hard for me to believe that something so light has an impact on the beam. But it does. And so does each step we take towards a vision of a peaceful, healthy planet. It's easy to feel overwhelmed by the magnitude of the problems we face right now—from hunger to war to global warming—and to question whether the little things we do make a difference. But when I get discouraged and wonder if we're making any progress at all and if it's even worth trying, I remember that fly and the words of Father Gerry.

Piti piti n a rive.

᪣

Margaret Trost is the founder and director of the What If? Foundation, a nonprofit organization dedicated to providing hope and opportunity to impoverished children in Haiti. Working in partnership with a Haitian community, the foundation supports food and education programs that make a difference in the lives of thousands of children. Margaret is the author of *On That Day, Everybody Ate: One Woman's Story of Hope and Possibility in Haiti* (Koa Books, 2008). For more information visit www.whatiffoundation.org and www.onthatdayeverybodyate.org.

The Grandmothers Speak

JENEANE PREVATT and ANN ROSENCRANZ

"We grandmothers have come from far and wide to speak the knowledge we hold inside. In many languages, we have been told it is time to make the right changes for our families, for the lands we love.... We are at the threshold. We are going to see change. If we can create the vision in our heart, it will spread. As bringers of light, we have no choice but to join together. As women of wisdom, we cannot be divided. When the condor meets the eagle, thunderbirds come home."

–Agnes Baker Pilgrim, oldest living female of the Takelma Band, Rogue River Indians, and keeper of the Takelma Sacred Salmon Ceremony, speaking at the inception of the International Council of Thirteen Indigenous Grandmothers in 2004.

There is a living legend, a story of awakening, calling us back in time to an original way of life and evoking memories that we humans carry deep in our cores. Around the planet many First Nation prophecies speak of the coming of a council of thirteen grandmothers as a catalyst and a calling to awaken to the new world that is announcing itself. According to these prophecies, the grandmothers are actually calling it into our awareness.

We [the authors of this essay] share this story that our grandmothers and grandfathers shared with us, and that we in turn share around the fire with our grandchildren. They have told us not to fear; a time of great shifting is upon us. "Where there is life, there is hope," they say. Our elders have told us that when we surrender our need to know and then devote ourselves to the Unknown, the veils part and we participate in miraculous healings from terminal illness, psychosis, and chronic depression. The original people pass on these ways of prayer in which all life is honored as sacred. When entering into dialogue with Creation's intelligence, we reverently approach life as a mystery and it responds to us. When we live without expectation and judgment, we accept what appears. We wake up in the dream and begin to participate in its weaving. In this weaving, all aspects are essential. Challenges stimulate solutions. And those solutions begin to create the new dream.

In following this dialogue with Creation, our community (called Kayumari) has been led to indigenous elders who encouraged and developed this way of seeing. Over the last twenty years, our elders took us into their families and shared their prophecies of this time of purification. They tell us that we are living in an era of upheaval, of choice and great healing. The nature of life brings all into balance. We have come to a time where the pendulum of balance has swung to an extreme, and our task is to participate in the rebalancing.

In their sacred relation to the earth, the original peoples take care of Her, serve Her, and know Her as their mother. We listened as elders spoke to us of the gathering of sacred clans, the unity across religious, racial, and gender boundaries. This challenging era demands that we unite as one people and fulfill our birthright to sustain our relation to one another and to our Mother Earth.

Our community heard the call. In November 2003 the members of Kayumari's nonprofit church, the Center for Sacred Studies (CSS), sent an invitation out to indigenous grandmothers around the world asking that they consider joining in council together. We explained that we needed their voice; the world needs the voice of the Life Givers. We wanted to help their voices be heard in a world tipped out of balance and disconnected from the sources of life.

This vision was born out of our international community (Kayumari), originally based in the Sierra Foothills in California. (We recently entered into a collaboration with a Buddhist retreat center on Black Mountain, located in the North Bay Area of San Francisco.) In Kayamuri's formative days in 1995, people from fourteen countries and many First Nations peoples gathered together to hold the dream of a safe place to pray. Since that time the community has continued to grow as an extended family of prayer and to touch many others around the world. We have grown a church together, the Center for Sacred Studies. We have prayed together and been in sacred relation with each other; holding each other as we birthed our babies; holding each other in song and prayer as some of us, young and old, died before us. When our sisters died, we adopted their children. We have nursed each other's babies. As our elders inspired us to do, we have dedicated ourselves to the regeneration of sacred culture.

On October 13, 2004, thirteen remarkable grandmothers answered the CSS's call. They gathered at the Menla Retreat Center in upstate New York from the mountains of Tibet and Nepal, the tundra of Alaska, the Brazilian Amazon, the highlands of Central America, the plains of North America, the Black Hills of South Dakota, the mountains of Oaxaca, the great forest of the American Northwest, the desert of the American Southwest, and the rainforest of Central Africa.

At this historical meeting they established a global alliance, the International Council of Thirteen Indigenous Grandmothers. They shared the visions and prophecies that guided them to say "yes" to joining our resources and prayers dedicated to a global movement for world peace and unity. Grandmother Rita Blumenstein, a seventy-five-year-old Yupik traditional healer, was guided by her great-grandmother. In 1942, when she was nine years old, her great-grandmother gave her thirteen stones and thirteen eagle plumes, telling her that she would be part of a council of thirteen and to save these precious relics for that time. At this first gathering in 2004, the day after declaring themselves an alliance, Grandmother Rita passed these thirteen stones and feathers to each and every grandmother around the table and took one for herself. In that moment she knew that her Grandmother stood behind her with all her ancestors, and that the time they had been preparing her for had arrived.

Grandmother Rita told us that Mother Earth is angry. She wants to shake us awake. She wants us to come back to ourselves. She wants us to realize that we are creating this reality. Will we react to this crisis with rage, hopelessness, and fear, or will we respond with love and prayer, approaching Creation for an answer?

The original peoples of this planet are the antidotes of our troubled time. Their traditional ways have sustained life since its beginnings. As a cultural presence the people have withstood innumerable planetary upheavals, war, genocide, and famine. Humanity has now arrived at the original ground. Every possibility that humankind has ever dreamed (abundance, war, disease, healing) is on the table of choice right now. It is imperative that we wake up from this nightmare, claim the choice, and take responsibility to rebalance our world.

Takelma Siletz Grandmother Agnes Baker Pilgrim tells us that the most important journey for us to embark on is the eighteen-inch journey from our head to our heart. The Grandmothers on the Council are connected to this sacred heart, for they are Life Givers, united with the Grandmother of all Life, the one without ancestors, who nourishes us without question.

The Grandmothers Council is not alone. Elders are gathering all over the earth. The International Council of Thirteen Indigenous Grandmothers is one of the many instruments of the Grandmother of all Life. They have united to renew our hope. From the four corners of the world the Grandmothers call to us, "Wake up, wake up! The time is *now.*"

RESOURCES

www.grandmotherscouncil.com
www.forthenext7generations.com
www.sacredstudies.org (for more on intentional community)
www.cssministryofprayer.org

❧

Kayumari is an international community in operation since 1995, currently based in Black Mountain Retreat, Cazadero, California. In 1998 the vision for the Grandmother Project started to catalyze their teachings and led to eventual creation of the International Council of Thirteen

Indigenous Grandmothers in 2004. In 2002 the Center for Sacred Studies (CSS) was formally recognized as a 501(c)3 nonprofit organization, and it currently serves as the umbrella group for the Grandmothers Council. In its alliance with the Grandmothers Council, the CSS continues to raise the funds and organize their Council Gatherings as well as facilitating intra-council communication and outreach to the world.

Jyoti (Jeneane Prevatt, PhD) is an internationally renowned spiritual advisor with a PhD in Transpersonal Psychology, which includes two-and-a-half years of postgraduate study at the C.G. Jung Institute in Zurich, Switzerland. She is one of the founders of Kayumari, one of the conveners of the International Council of Thirteen Indigenous Grandmothers (www.grandmotherscouncil.com), and the Spiritual Director of the Center for Sacred Studies (www.sacredstudies.org). She travels the world and is dedicated to bringing unity to the planet by developing alliances between the guardians of indigenous culture and traditional medicine. Her Web site is at www.cssministryof prayer.org.

Ann Rosencranz, MA, is an ordained minister, translator, singer, and writer, holding an MA in Philosophy and Religion. Ann is one of the conveners of the International Council of Thirteen Indigenous Grandmothers and the Program Director for the Center for Sacred Studies Grandmothers Program. Ann oversees all the operational and developmental responsibilities of the Council and travels with them nationally and internationally to walk a prayer of peace and cultivate alliances between indigenous and non-indigenous peoples.

The Ultimate
Miracle Worker

JALAJA BONHEIM

How shall we live in these times? How shall we respond to the awareness of what is happening in our world?

Perhaps the first step is to acknowledge honestly: *we don't know.* We may have our guidelines, our intentions, our beliefs about what is helpful and what is not. But do any of us see a clear path from the mess we've created on our planet to a peaceful sustainable world? No, we don't.

We don't, because we *can't.* The process we're involved in today is not one that the human intellect is capable of penetrating. Like the dance of subatomic particles, it's multidimensional, mysterious, and impossible for our minds to grasp.

Once we come to grips with this fact, we can let go of the crippling assumption that we *should* know what to do—or that, at the very least, *someone* should know. Since nobody does, this can leave us feeling panicky and overwhelmed.

I believe that solutions *do* exist to our problems. But will we find them? That's another question. As Einstein said, problems can never be resolved at the level at which they were created. The environmental crisis was caused by the human mind—or more accurately, by the ways in which we habitually use it. But problems created by the mind cannot be resolved by the mind.

We are used to turning to the mind for guidance, and when it can't make out the path, we tend to feel hopeless. Yet if we can make peace with the fact that our mind is not in control of this journey, then we can open to the possibility of what some might call a miracle. We usually think of miracles as events that contradict the laws of nature, as when Jesus turned water into wine. But the kind of miracle I'm talking about

265

here does not contradict nature. Rather, it's guided by the intelligence of nature herself, who is the ultimate miracle worker.

Of course, life is a continuous miracle, in the face of which we can't help but bow down in gratitude and awe. Yet among all the feats of natural magic, one of the most extraordinary is surely the transformation of caterpillars into butterflies. You probably already know that when a caterpillar is ready to shape-shift, it forms a cocoon. But did you know that within that cocoon, it quite literally liquefies? It dies, and dissolves into a mass of separate cells.

Then, however, something truly amazing happens. Within that cellular goo, some of the old caterpillar cells begin to mutate into what biology calls *imaginal cells*—imaginal, because they carry within them the image of the butterfly-to-be. Nobody can predict which cells are going to transform, nor do we know what triggers the process.

Yet no sooner do the imaginal cells begin to appear than they come under attack from the old caterpillar cells. It's not hard to sympathize with the old cells—presumably, they feel they've been invaded by aliens. And so, they go on the offensive.

Interestingly, the imaginal cells don't even bother fighting back— they're far too busy working on their crazy butterfly project. Nonetheless, in the end, they emerge victorious. Some die, but most survive, and continue on their way, driven by their overwhelming desire to experience life in a butterfly body.

Do they know how to go about accomplishing this? Absolutely not. They don't have a clue. But they *do* know how to attune themselves to nature's intelligence, and let themselves be guided by her. And fortunately, nature knows exactly how to make butterflies. She's been doing it for a long, long time, and she's got it down to a T. However, the great work cannot begin until the imaginal cells connect. Isolated and alone as they initially are, they can achieve nothing. And so, they begin to reach out tendrils.

"I'm feeling lonely," we might imagine them saying. "Is there anyone else out there?"

And to their immense relief, there is. In fact, by the time they reach out, there are millions of other imaginal cells out there. And so, they connect, and together, they begin to weave the matrix out of which one day a butterfly will emerge—a gossamer speck of beauty dancing on the breeze.

We too are imaginal cells, weavers of a new world. Listening to the news, the idea of creating a peaceful, sustainable human civilization might seem like a mad fantasy. There are far too many challenges, all of them serious and potentially devastating. Like the World Trade Center, old structures are crumbling all around us. Yet like the imaginal cells, we too are many—far more than most of us realize. And in recent years, we too have been connecting, and have begun to weave the matrix of a new world.

Indeed, the creation of this book, this interweaving of our collective wisdom, is a perfect example of how imaginal cells operate. No one person could have created it. Nobody could have predicted what insights, ideas, images and stories it would contain. Yet here we are, holding it in our hands.

Imaginal cells aren't given an instruction manual. Nobody tells them what to do. Rather, they are guided from within. We too, can only attune ourselves to the vast consciousness that created our cosmos by quieting ourselves, turning inwards, and listening.

I am not referring to meditation, although meditation certainly is a powerful tool for detaching from the mind. Rather, I am speaking, quite simply, of *listening*. Let us turn to the source of guidance within, ask our questions, from the most mundane to the cosmic, and listen to the responses that rise up. I have witnessed thousands of people as they did this and discovered, sometimes to their great surprise, that insofar as they were truly willing to listen, they were gifted with insights and guidance that their conscious mind had no access to.

Can we, like the imaginal cells, make ourselves available to serve as agents of nature's infinite wisdom? If so, we may yet enable a planetary transformation no less miraculous than that of a caterpillar into a butterfly.

꧁

Jalaja Bonheim, PhD, is a public speaker, circle leader, and leadership trainer who has spent the last decades sharing her profound knowledge on how we can utilize circle gatherings to heal ourselves and our communities. She's the founder and visionary director of the Institute for Circlework, which empowers leaders from around the world, especially conflict zones, to serve as agents of peace, and the author of five books, including *Aphrodite's Daughters: Women's Sexual Stories and the Journey of the Soul* and *Living in Peace: A Vision of Hope for Humanity*. Visit the Institute for Circlework online at www.instituteforcirclework.org.

The Challenge
of Building Sustainably

SCOTT RODWIN

Perhaps I am naïve, because I *love* a challenge. "The situation is dire and we don't know how to solve it! We're almost out of time. The fate of the whole world is at stake!" Perfect. That's what makes an exciting and heroic story—and a challenge worth devoting my life to.

As an architect I have spent most of my adult life learning how to creatively solve problems. Designing a physical environment that supports life instead of destroying it—what tougher and more important problem could I choose to tackle?

Noteworthy facts:

1. In the U.S., buildings consume roughly 65 percent of our electricity and 30 percent of our raw materials; and they generate 30 percent of our waste and greenhouse gases. They are the largest single-sector impact on our natural environment.

2. The average American spends roughly 90 percent of his or her time indoors. The buildings we create are among the largest environmental shapers of the human experience, greatly affecting both physical and psychological health.

In 1990, my fourth year of architecture school at Cornell, I became part of a remarkable group called EcoVillage of Ithaca (NY), a sprouting "intentional community" (a term for a type of residential development designed to promote interaction and cooperation among neighbors). Two amazing women founded the group, one more the visionary, one more the pragmatic problem-solver. Together they created a dynamic and successful process that ultimately gave birth to a large and vibrant sustainable community. They showed me how both of those roles contribute to create a powerful new solution for housing ourselves.

After graduation, I moved to Colorado and lived at the Nyland Cohousing community in Lafayette for four years. Its forty-two town-homes are grouped on eight acres of the forty-two-acre rural development, with the rest of the land left undeveloped. The houses themselves are small and energy-efficient, passive-solar duplexes and triplexes, with traditional front porches that facilitate impromptu socializing. A large Common House (clubhouse) sits in the center, and community members have the option of eating home-cooked group meals there a few times a week. The central building also houses a library, fitness and rec area, guest rooms, teen and kids' play rooms, and a laundry area. Cars are kept to the perimeter of the property, and herds of kids safely run amok on the pedestrian pathways that tie the neighborhood together. There is a passive-solar greenhouse, a well-equipped workshop, organic gardens, play structures, and fields. It's colorful, rural, and a bit funky. The landscape is xeric (featuring drought-tolerant native plants for low water consumption) and employs permaculture techniques.

The community is self-managed and maintained by the residents, and all decisions are made by consensus. Despite the utopian-sounding program, the Nyland Cohousing project is organized on a conventional condominium/HOA (homeowner association) model and largely functions like an old-fashioned neighborhood. People have mortgages, regular families, and normal jobs. It was a great place to live.

At the same time, I was beginning my architectural career and in the process of co-founding another cohousing group, one that deliberately was located in town. Nomad Cohousing, where I have now lived for ten years, is a block away from a neighborhood market, cleaners, coffee shop, and bus stop. Our little eleven-unit town-home project was built as infill in an existing neighborhood. We are all clustered around a small courtyard and share our Common House with the live theater next door. In the morning, I walk out to the lush courtyard with my breakfast and sit with my neighbors chatting about our lives and current events. I have a little private backyard, and my best friend lives thirty feet away. My total utility bill is about $50 a month, and our HOA fees are equally small because we maintain the facility ourselves. It works and I love it.

Now don't get me wrong. It took a lot of work to create this: hundreds of hours of organizational and design meetings; compromise, patience, perseverance, tolerance, surrender, humility, and compassion in learning how to be flexible and how to live with other people. Visioning a goal for how we would like to live. Being creative in solving hundreds of small questions like "Can we have only one lawn-mower for eleven households?"

If you want to take on the challenge, you're going to have to work for it. And as far as I'm concerned, that makes for a good life. If it were easy, I wouldn't be forced to grow. Yes, it is harder (up front). It does take more effort. You do have to be extra creative, committed, and intelligent about how you live your life. And I choose to do it.

My clients come to me specifically because my firm is known for green design. We will work on anything—houses, schools, churches, offices, restaurants. I don't believe there is a bad project. I would design a Walmart if they asked me. Why? Because I can make it as good as it can be. And the worse is it to begin with, the more opportunity there is to make it wonderful.

Ironically, Walmart, often cited as the epitome of environmentally evil business, is currently embarking on one of the largest and most complete green building ventures in the world. The only entity going bigger is China. The thing they have in common is their architect. No, not me. It's a staggeringly inspiring fellow named William McDonough (co-author of the revolutionary book on responsible consumerism called *Cradle to Cradle*). He is the former Dean of the Architecture School of the University of Virginia, and his work has inspired a lot of what I do. In 1993 McDonough gave a speech to the American Institute of Architects here in Colorado. It was the first time I had ever heard anyone speak publicly about the moral imperative of sustainable design. Like the other four hundred people in the audience, I leapt out of my chair (I think I actually stood on it) and gave him a five-minute standing ovation at the conclusion. I got the gospel.

McDonough has few built projects to his name, and he is rarely the designer of the buildings he works on. He landed clients like Ford, Walmart, and China not because of a pretty portfolio, but because he was able to convince them that they could go green or they could go the way of the dinosaur. Did they change their ways out of the goodness of their hearts? Mmmmmm ... who knows? But we can bet on the fact that survival is what inspired them to immediate action. What action? China hired McDonough to design a dozen completely sustainable, new prototype cities. Ford and Walmart are both undertaking worldwide green building initiatives of a colossal scale.

What did McDonough say that caused this quantum shift? Foremost, he showed the Chinese government and these transnational corporations that green is good for the long-term bottom line. For example, if you spend a bit more up front on an energy-efficient mechanical system, you will make that money back in energy savings each year. If you specify the "more expensive" non-toxic paints, natural ventilation, and good solar day-lighting in your office, you will have fewer sick days among employees, less staff turn-over, higher sales, and significantly better worker productivity (as governmental and private industry studies have demonstrated). And there are those indirect things that have a less obvious relationship: nuclear power seemed pretty cheap at first, but it doesn't look quite so clever from a business perspective now that we've had to spend hundreds of billions of taxpayer dollars for dubious cleanup and long-term containment sites. Even the utility companies now say that conservation is the most cost-efficient way to maximize available energy and reduce emissions. Whether we're looking at pollution, global warming, deforestation, water contamination, mining, and timber harvesting impacts, or general resource depletion, I see a growing consumer awareness that how we personally choose to live has a public and global impact. Both individuals and businesses are coming to understand that environmental sustainability is a good investment. This gives me hope.

What's the good of a fine house if you don't have a tolerable planet to put it on?
  –Henry David Thoreau

Sometimes when I see people doing amazing things, I feel pretty darn small and my efforts look virtually insignificant. But then I remember the starfish story. You've probably heard it.

A man is walking along a beach after a big storm. There are tens of thousands of starfish washed up and dying. In the distance, a small boy is picking them up and throwing them one by one back into the ocean. The man walks up to him and says, "What are you doing?" "Saving them," replies the boy. "You're crazy. There are thousands and thousands of them. You can't possibly make a difference." The boy was quiet for a moment then looked down, picked one up, and threw it in. "Made a difference to that one," he said.

My clients are unusually well educated about the environment and do not yet represent the majority, but this segment of the population is growing as the building trend toward sustainable design expands. People want healthy, energy-efficient, durable, non-toxic buildings. They want to feel good about their choices.

One of my favorite quotes is from Winston Churchill: "You can always count on Americans to do the right thing . . . after they have tried everything else." This can be applied to our building pattern. I am a bit of an optimist. I believe that America, and the world, has the energy, ability, and creativity to get ourselves out of the mess we have created. When I figure out how to get solar panels into a school for the same price as a conventional boiler, I feel a sense of accomplishment. When I sell a client on bamboo flooring and strawbale walls, I know I have saved part of a forest that day. When I show a developer how to make money and conserve land at the same time with a community-fostering cohousing site plan, I have enrolled someone in a better possibility. This inspires me.

I believe we create a better future by learning from our past and bringing back those elements that worked well, but we must also invent new ways of living to solve the problems that remain. McDonough's innovative two-track approach to this issue is illuminating. He separates the built world into two groups: biological nutrients (natural things) and technical nutrients (man-made things). He argues that it makes sense to keep these two categories separate. That means that biological products like wood and paper should be recycled into other products of that type (wood into other wood products, for example), while man-made things like steel, glass, concrete, and plastic should be recycled into more steel, glass, etc. In McDonough's definition, it is critical that recycling not mean "downgrading." If trees become toilet paper and that becomes mixed landfill, the resource has been downgraded to a waste product and its value is destroyed. It has to stay in the cycle and preferably at a similar level to the original product. This is how we begin to contain our waste stream and resource consumption.

Former President Bush invited McDonough to the White House to advise his administration. Why? Perhaps because McDonough advocates the eventual elimination of all environmental regulations—okay, that's a head-scratcher. His argument is that given a true free-market economy (one in which oil is not subsidized as it is now), the simple, green, nontoxic, and local will beat the pants off big-box imported junk. And that philosophy trickles down to every aspect of the physical environment. If unsubsidized gasoline is $7 a gallon, people will be more likely to design homes that are sized appropriately for their real needs. The homes will be passively heated and cooled and tuned to a specific climate and site. They will be built to last, be flexible, and be recyclable, located close to necessary services and community. They will be designed, built, and disposed of with care.

Each project I do deserves this attention and passion for the challenge. When I graduated from school I knew that I had a choice; the green architecture movement at that time represented only 1 percent of the profession. I could feel part of a self-righteous minority while rebelling against the profession from the outside, or I could become part of the professional organization American Institute of Architects (AIA) and help change it from within. I chose the latter and in doing so joined a small wave of designers committed to rediscovering a balance and symbiotic relationship between the man-made and the natural environment. Our goal was not just to be less bad, it was to find a way to make buildings good, even regenerative. We are still a long way from that goal, but we are racing faster than anyone could have imagined in the right direction.

About a decade ago, the AIA changed its charter to include ecological sustainability as one of its principal tenets. There is not an AIA design award in the country that now fails to make it a priority. The AIA and virtually every other allied professional group (including builders, developers, planners, landscape architects, engineers, and interior designers) are waging an enormous education campaign to help research and promote sustainability. And it has had an effect. The U.S. Green Building Council has exploded onto the national scene with a green certification program called LEED (Leadership in Environment and Energy). It has proven wildly popular, with the number of projects using it increasing exponentially every year since the program's inception.

What is a green building? Does it really make a difference?

A typical American home creates about thirty thousand pounds of carbon dioxide per year, which directly contributes to global warming. Additionally, carbon dioxide emissions have a general correlation to a host of other environmental impacts including pollution, resource consumption, and environmental degradation. Consider a house or commercial building that is carbon-neutral, meaning one that is "net-zero" for energy use and carbon dioxide emissions, putting as much back into the system as it takes out. This would be one of the key elements of a truly sustainable world.

Is it possible?

We're doing it right now.

We recently completed a LEED Platinum, net-zero, carbon-neutral home. But would an average American really want that? Doesn't it mean that we would have to live in teepees or in desert caves made of recycled tires? Not anymore. This house is beautiful, modern, full-sized, and mainstream. It has geothermal heat and cooling, solar electric panels and solar hot water, super-insulated windows, walls, and roof, grey water recycling, native landscaping, natural day-lighting, non-toxic finishes, sustainably harvested lumber, and a host of recycled components. Modern energy-efficient buildings can be indistinguishable from their conventional counterparts.

Banks like loaning on energy-efficient buildings. Realtors like selling sunny, thermally comfortable houses with small utility bills. Businesses like the higher retail sales and greater employee production that green buildings support, which lead to higher tax revenues, which can help fund things like healthy schools. Administrators like telling parents that their kid's schools have great indoor air quality and natural day-lighting (which both contribute to lower absenteeism and substantially higher test scores). And parents like telling their neighbors about how their bamboo flooring saved a grove of trees that their better-educated and healthier kids will now have the pleasure of playing in.

Sustainable design is spreading everywhere and picking up speed. For a long time, we were pushing a boulder up a hill, but now we have passed the tipping point . . . it is unstoppable.

This has been one hell of a challenge. And there's still so much more to do.

❧

Scott Rodwin is the president of Rodwin Architecture and a green construction company, Skycastle Homes, in Boulder, Colorado. He was the American Institute of Architect's 2006 "Young Architect of the Year" for the Western Mountain Region and is the recipient of more than a dozen awards for excellence and leadership in sustainable design. He has helped design some of Colorado's greenest buildings, including homes, churches, restaurants, and schools. Scott is a frequent lecturer and teaches green building for the City and County of Boulder. His Web site is www.rodwinarch.com.

The Optimism
of Uncertainty

HOWARD ZINN

In this awful world where the efforts of caring people often pale in comparison to what is done by those who have power, how do I manage to stay involved and seemingly happy?

I am totally confident not that the world will get better, but that we should not give up the game before all the cards have been played. The metaphor is deliberate; life is a gamble. Not to play is to foreclose any chance of winning. To play, to act, is to create at least a possibility of changing the world.

There is a tendency to think that what we see in the present moment will continue. We forget how often we have been astonished by the sudden crumbling of institutions, by extraordinary changes in people's thoughts, by unexpected eruptions of rebellion against tyrannies, by the quick collapse of systems of power that seemed invincible.

What leaps out from the history of the past hundred years is its utter unpredictability. A revolution to overthrow the czar of Russia, in that most sluggish of semi-feudal empires, not only startled the most advanced imperial powers but took Lenin himself by surprise and sent him rushing by train to Petrograd. Who would have predicted the bizarre shifts of World War II—the Nazi-Soviet pact (those embarrassing photos of von Ribbentrop and Molotov shaking hands), and the German Army rolling through Russia, apparently invincible, causing colossal casualties, being turned back at the gates of Leningrad, on the western edge of Moscow, in the streets of Stalingrad, followed by the defeat of the German army, with Hitler huddled in his Berlin bunker, waiting to die?

And then the postwar world, taking a shape no one could have drawn in advance: the Chinese Communist revolution, the tumultuous and violent

Cultural Revolution, and then another turnabout, with post-Mao China renouncing its most fervently held ideas and institutions, making overtures to the West, cuddling up to capitalist enterprise, perplexing everyone.

No one foresaw the disintegration of the old Western empires happening so quickly after the war, or the odd array of societies that would be created in the newly independent nations, from the benign village socialism of Nyerere's Tanzania to the madness of Idi Amin's adjacent Uganda. Spain became an astonishment. I recall a veteran of the Abraham Lincoln Brigade telling me that he could not imagine Spanish Fascism being overthrown without another bloody war. But after Franco was gone, a parliamentary democracy came into being, open to Socialists, Communists, anarchists, everyone.

The end of World War II left two superpowers with their respective spheres of influence and control, vying for military and political power. Yet they were unable to control events, even in those parts of the world considered to be their respective spheres of influence. The failure of the Soviet Union to have its way in Afghanistan, its decision to withdraw after almost a decade of ugly intervention, was the most striking evidence that even the possession of thermonuclear weapons does not guarantee domination over a determined population.

The United States has faced the same reality. It waged a full-scale war in Indochina, conducting the most brutal bombardment of a tiny peninsula in world history, and yet was forced to withdraw. In the headlines every day we see other instances of the failure of the presumably powerful over the presumably powerless, as in Brazil, where a grassroots movement of workers and the poor elected a new president pledged to fight destructive corporate power.

Looking at this catalogue of huge surprises, it's clear that the struggle for justice should never be abandoned because of the apparent overwhelming power of those who have the guns and the money and who seem invincible in their determination to hold on to it. That apparent power has, again and again, proved vulnerable to human qualities less measurable than bombs and dollars: moral fervor, determination, unity, organization, sacrifice, wit, ingenuity, courage, patience—whether by blacks in Alabama and South Africa; peasants in El Salvador, Nicaragua and Vietnam; or workers and intellectuals in Poland, Hungary, and the Soviet Union itself. No cold calculation of the balance of power need deter people who are persuaded that their cause is just.

I have tried hard to match my friends in their pessimism about the world (is it just my friends?), but I keep encountering people who, in spite of all the evidence of terrible things happening everywhere, give me hope. Especially young people, in whom the future rests. Wherever I go, I find such people. And beyond the handful of activists there seem to be hundreds, thousands, more who are open to unorthodox ideas. But they tend not to know of one another's existence, and so, while they persist, they do so with the desperate patience of Sisyphus endlessly pushing that boulder up the mountain. I try to tell each group that it is not alone, and that the very people who are disheartened by the absence of a national movement are themselves proof of the potential for such a movement.

Revolutionary change does not come as one cataclysmic moment (beware of such moments!) but as an endless succession of surprises, moving zigzag toward a more decent society. We don't have to engage in grand, heroic actions to participate in the process of change. Small acts, when multiplied by millions of people, can transform the world.

Even when we don't "win," there is fun and fulfillment in the fact that we have been involved, with other good people, in something worthwhile. We need hope. An optimist isn't necessarily a blithe, slightly sappy whistler in the dark of our time. To be hopeful in bad times is not just foolishly romantic. It is based on the fact that human history is a history not only of cruelty but also of compassion, sacrifice, courage, kindness.

What we choose to emphasize in this complex history will determine our lives. If we see only the worst, it destroys our capacity to do something. If we remember those times and places—and there are so many—where people have behaved magnificently, this gives us the energy to act, and at least the possibility of sending this spinning top of a world in a different direction. And if we do act, in however small a way, we don't have to wait for some grand utopian future. The future is an infinite succession of presents, and to live now as we think human beings should live, in defiance of all that is bad around us, is itself a marvelous victory.

(from *A Power Governments Cannot Suppress*)

≫

Howard Zinn (1922–2010) was a shipyard worker and an Air Force bombardier before he went to college under the G.I. Bill of Rights, getting his doctorate at Columbia University. He taught at Spelman College in Atlanta for seven years, becoming involved in the Southern civil rights movement. Moving north, he taught at Boston University and was active in the movement against the war in Vietnam. He wrote many books, the best known of which is *A People's History of the United States*. For more about Howard Zinn, visit www.howardzinn.org.

Afterword

Sabbaths: VI

WENDELL BERRY

It is hard to have hope. It is harder as you grow old,
for hope must not depend on feeling good
and there is the dream of loneliness at absolute midnight.
You also have withdrawn belief in the present reality
of the future, which surely will surprise us,
and hope is harder when it cannot come by prediction
any more than by wishing. But stop dithering.
The young ask the old to hope. What will you tell them?
Tell them at least what you say to yourself.

Because we have not made our lives to fit
our places, the forests are ruined, the fields eroded,
the streams polluted, the mountains overturned. Hope
then to belong to your place by your own knowledge
of what it is that no other place is, and by
your caring for it as you care for no other place, this
place that you belong to though it is not yours,
for it was from the beginning and will be to the end.

Belong to your place by knowledge of the others who are
your neighbors in it: the old man, sick and poor,
who comes like a heron to fish in the creek,
and the fish in the creek, and the heron who manlike
fishes for the fish in the creek, and the birds who sing
in the trees in the silence of the fisherman
and the heron, and the trees that keep the land
they stand upon as we too must keep it, or die.

This knowledge cannot be taken from you by power
or by wealth. It will stop your ears to the powerful
when they ask for your faith, and the wealthy
when they ask for your land and your work.
Answer with knowledge of the others who are here
and of how to be here with them. By this knowledge
make the sense you need to make. By it stand
in the dignity of good sense, whatever may follow.

Speak to your fellow humans as your place
has taught you to speak, as it has spoken to you.
Speak its dialect as your old compatriots spoke it
before they had heard a radio. Speak
publicly what cannot be taught or learned in public.

Listen privately, silently to the voices that rise up
from the pages of books and from your own heart.
Be still and listen to the voices that belong
to the streambanks and the trees and the open fields.
There are songs and sayings that belong to this place,
by which it speaks for itself and no other.

Found your hope, then, on the ground under your feet.
Your hope of Heaven, let it rest on the ground
underfoot. Be lighted by the light that falls
freely upon it after the darkness of the nights
and the darkness of our ignorance and madness.
Let it be lighted also by the light that is within you,
which is the light of imagination. By it you see
the likeness of people in other places to yourself
in your place. It lights invariable the need for care
towards other people, other creatures, in other places
as you would ask them for care toward your place and you.

No place at last is better than the world. The world
is not better than its places. Its places at last
are no better than their people while their people
continue in them. When the people make
dark the light within them, the world darkens.

☙

Wendell Berry is farmer, essayist, conservationist, novelist, teacher, and
poet. He has been the recipient of numerous awards and honors and is
the author of more than forty books of poetry, fictions, and essays. Berry
has farmed a hillside in his native Henry County, Kentucky, for more
than forty years. His Web site is www.wendellberrybooks.com.

Gratitude

My days and nights of the past four years have been in service to this anthology. *Hope Beneath Our Feet* seems to have chosen me, and we have been constant companions. As this book goes into the world I take joy in naming the people who have stepped forward to help along the way.

The editorial boards at publishing houses big and small were excited about the authors and ideas in this compilation. Then, repeatedly, the marketing departments would nix acquiring the book. I'm extremely grateful for the exceptional team at North Atlantic Books; this publishing house cares about ideas and people first. Richard Grossinger recognized that this book is "remarkably cogent, fierce, and intelligent." Erin Wiegand shepherded this anthology through all the stages of publishing. Kathy Glass, who lives off the grid in a straw bale home and gives her time to save old-growth forests on the North Coast of California, diligently worked with the authors to edit their writings. Kudos go to Robin Donovan for copyediting and to Kat Engh for getting word of this anthology out to the world.

Many essays were submitted that will have to wait for future volumes. This project allowed me to sleep better when I found out that there are so many people who care enough to act.

A diverse group of individuals offered their time as "first-impression readers." As the essays poured in, they offered their gut reactions and ideas for improvement: Alex Chasin, Anne Aronov, Barry Shapiro, Brando Brandes, Catherine Musinsky, Dan Zola, Deborah Feld, Deborah Watrous, Diane Cella, Elizabeth Alach, Gene Broadway, Jill Cooper, Jovanina Pagano, Lolita Valianos, Lori Thomson-Sweet, Mars Miquelon, Max Gautier, Megan Orwig, Nancy Shapiro, Neige Christenson, Owen Jones, Peggy Holcomb, Pen Dale, Peter Rosselli, Rachel Hollowgrass, Rhonda Morton, and Susan Singer.

Special thanks go to a few who went the extra distance: to the activist, Vivienne Simon; to the man who brings poetry and music into his dissent,

Shepherd Bliss; for ongoing support of my book projects, Kristelle Bach; to the weavers of people, Bob Banner and Tzeporah Berman; and to the writer's midwife, Elianne Obadia. Thanks to Deborah Watrous for our conversations and the book retreat at Crooked Pond. And an immense hug to the one who was there every step of the way with insight and honest feedback, Susan Lehotsky.

My respect goes to the staff and volunteers who bring vitality to the Earthdance Retreat center. Thank you especially to Margit Galanter and Spirit Joseph for your generosity in presenting me the secluded Claudio's cabin for so many book retreats that I lost count. This anthology would not be in our hands if it were not for your trust and support.

To my dance companions around the world who have such a small carbon footprint because what is important is not what they acquire but the life they live: You create a refuge from the sorrows of the world. Now get out there and get active; you can't live in the refuge forever.

To my siblings Nina and Peter Keogh, for upholding art as a doorway to a meaningful life.

I'm grateful to the incredible force of nature that is my son, Dylan, and to my stepsons, Wyatt, William, and Stefan; you are my great teachers. Seeing you growing up gives me the impetus to get off my ass to *do* something.

I'm grateful for the walks through the Mexican countryside with my first mentor, Pierre Delattre, and to Robert Bly, who showed me that I could fall in love with language.

My connections to people give me the greatest ground for thanksgiving. This includes my students, who demand that I stay vital in my body and imagination, and also the champions of the dance from Contact Improvisation who host me to teach what I love all over the world. There are individuals in our lives who believe in us, and it is on the pillar of that belief that we can trust ourselves to do something greater. Some of these friends in my life include: Anne Aronov, Dan Zola, Don Mill, Karen Roeper, Leigh Hollowgrass, Mary Ford, Michael Steinberg, Nina Bindi, Owen Jones, Peter Rosselli, and Valerie Mejer.

My heartfelt appreciation goes to the authors in this volume: you inspire hope by your generosity, action, and caring. You are doing work in the world on behalf of all of us.

Finally, I extend gratitude beyond words to my companion, lover, and wife, Liza—my life is dedicated to you.

–Martin Keogh
Easton, Massachusetts, 2010

Index

A

Ableman, Michael, 120–21
abolitionists, 5
Ackerman, Diane, xiv, 144
activism, xiv (*see also* ecology;
politics); and the absence of a
national movement, 279; and a balance of hope and complacency, 12;
and bargaining, 227; and a big
vision broken down into small
steps, 258; channeling despair and
anger into, 197–98, 226–27, 229;
and choice, 9, 31, 76, 77, 79, 80, 81;
and the difference between issues
and entry points, 77, 78, 79;
environmental, 4–5, 63–64, 194,
279; and ethical reason, 194–95;
and excuses, 132; and fear, 69, 228,
229; five-hundred-year plan, 93–95;
and gratitude, xiv; inner work of,
184–87; mobilization required, 69;
outer work of, 184, 186; and the
personal as political, 187, 199, 229,
248; and political involvement,
122–24; and predicting outcomes,
79–80, 277, 279–80; simple tactics,
69–70; and social revolution, 62,
64–65, 66, 130–31; staying inspired,
193–96; teachers of, 193; and
thinking ecologically, 250–51; three
types of activities, 202–5; and
voting, 87–88; and wonder, 178
adolescence: and hormones, 24–25; of
humanity, 24, 25–28, 230
advertising, 162–63
affinity groups, 251
agribusiness, 66, 166–67
ahimsa, 168
AIA. *See* American Institute of
Architects
AIDS, 54, 76

Airplane Fast, 41
Alach, Kristine, 92
Alexander Technique, 164
alternative medicine, 63, 205, 229
Amazon rainforest. *See* rainforests
American Institute of Architects
(AIA), 274–75
Anarctic ice, 80, 112, 142 (*see also*
Arctic ice)
anxiety (*see also* stress); eco-anxiety,
58–59, 201, 203; as environmental
toxin, 226–27; questioning
thoughts that cause, 189; reactions
to, 58–59
apartheid, 254
Apollo, 145, 148
Arctic ice, 30, 80, 99, 100 (*see also*
Anarctic ice)
Ardagh, Chameli, 139
Aribabi, Rancho El, 235–36
ark, 45–46
Armstrong, Jeanette, 152
arts, 151, 153–54, 155–56 (*see also*
creativity; writing)
Auschwitz, 18
Avila, Sergio, 235, 236, 237
Awakening Women Institute, 139

B

Bali, 209–12
Barks, Coleman, 291
Bartlett, Susan, 124
baseball, 161
BCollective, 157
Bean Keepers, 94–95
Beauchamp, Father Bill, 7
Begley, Ed Jr., 30

Benyus, Janine, 5
Berman, Tzeporah, 199–200
Berry, Wendell, x, 285; and the cheap-energy mind, 33; civil disobedience on climate change, 100; on environmental crisis as a crisis of character, 30–31; on growing food, 36; on specialization, 32
Beyond Organic, 121
Bhutan, 186
bicycles, 73–75
biodiversity, 255
biology, 156
Bioneers, 119, 126, 224
bioremediation, 126, 130
biosphere, 49
Blumenstein, Rita, 262
Bly, Robert, 291
Bodhi Be, 242
body: advice for, 163–64; as community, 6; disconnection and dissociation from, 150, 154, 160, 163; and the earth, 55, 56, 145, 152, 153, 154; and ecology, 156, 162; and the embodied arts, 150; felt-sense of (*see* soma); living in an electronic world, 159–60; and localism, 150–51; as lower animal part, 159; mammals sharing our DNA, 145; and physical education, 160–63; re-inhabiting, as activist strategy, 150, 154, 158; and stress, 162; as a tool for spiritual communication, 158, 159
bodywork, 164
Bonheim, Jalaja, 267
Bookchin, Bea, 193, 194
Bookchin, Murray, 193, 194, 195, 250
Bounoua, Lahouari, 16
breath, 170, 171–72, 174
Breen, Bill, 110
Brower, David, 115
Brown, Molly Young, 186
Bucky. *See* Fuller, Buckminster

Buddha, 137, 190
Buddhism, 213
buildings, 269; green, 250, 271, 272, 273, 274, 275–76; natural, 250
Bush, George, 274
business, 106–8, 109
butterflies, 265–66
Buzzell, Linda, 206

C

Caldicott, Helen, 227
Campbell, Joseph, 291
Cape Leopard Trust, 149
carbon, 33, 99
carbon dioxide, 49–50, 275
carbon footprint, xii, 29, 30, 50, 228; and food, 30, 35, 81; and gardening, 35–36
Carson, Rachel, 178, 226
Carter, Jimmy, 29, 34
catastrophe. *See* environmental catastrophe; social catastrophe
caterpillars, 265, 267
Center for a New American Dream, 48
Center for Sacred Studies (CSS), 261, 262, 264
Center for Urban Agriculture, 117, 120
Chalquist, Craig, 206
chanting, 163
cheap-energy mind, 33, 35, 36
children, 76, 80, 83, 161, 164, 180; Haitian, feeding, 256–58
China, 29, 34, 271, 272, 277–78
choice, 31, 54, 263; planetary effects of, 76, 77, 79, 80, 81; and spiritual development, 8–9
Christenson, Neige, 85
Churchill, Winston, 273
Cirque Dirt, 95
civil disobedience, 66, 100
Clarkson, Thomas, 5

Class of 2009, 3–4, 7

climate change. *See* environmental catastrophe; global warming

Climate Conference in Copenhagen, 99, 168

"Club of Rome" report, 111

Cobb, John, 21

coffee, 77, 78

Cohen, Andrew, 10

Coming Back to Life (Macy and Brown), 186

commercialism. *See* consumerism

community: and the arts, 155; bonding when death occurs, 240; building local and global, 205; created by water, 236; disconnection and dissociation from, 154; fundamental issues for, 60; and the future, 42, 44–45; gardens, 35, 104; intentional, 63, 269–70; our longing for, 64; and specialization, 32–33; sustainable, 229–30

community-supported agriculture (CSA), xiii, 43, 63, 79

composting, 35, 36, 87, 88

ConAgra, 106

conservation, energy, 51 (*see also* activism); acting as if acting makes a difference, 34–35; and the bicycle, 73–75; and China, 29, 34; laws and money required, 30; and living locally, 41; movement, 4; and personal virtue, 29–31, 32; and the self-regulation of life, 50, 51; and Vicki Robins' letter from 2030, 42–43; viewed by utility companies, 272

consumerism: and Americans' inability to downsize, 62–63; and dissociation from our true sense of self, 162–63; as environmental toxin, 226–27; rebellion against, 64; and a Simplicity Circle, 204; and social and economic change, 63, 66, 248; as violence, 168

contemplation, 27

Conversation Cafes, 48

Copenhagen conference. *See* Climate Conference in Copenhagen

coral reefs, ix

courage, 80

Cousins, Norman, 46

Coyne, Kelly, xiv, 105

Cradle to Cradle (McDonough), 271–72

Crazy Horse, 238

Creation Spirituality, 203

creativity, 184, 186 (*see also* arts; writing)

Creator. *See* God; Goddess

CSA. *See* community-supported agriculture

CSIRO Wildlife Survey Section, 111

CSS. *See* Center for Sacred Studies

Cuba, 119–20

culture: death of dominant, 225, 230; disposable, 240; and dissociation from our true sense of self, 162–63; indigenous, 153, 156; killing the planet, 132–34; materialistic view imbedded in, 62; reactions to dying of, 225–29

Czechoslovakia, 34

D

Daly, Herman, 21

dance, 153–54

Dance and Human History, 161

Darwin, Charles, 6

DDT, 13

death: care, 240, 241–42; and denial, 225, 238; of dominant culture, 225–29; a good day for, 238–39; out of balance relationship with, 239, 240; psychological responses to, 225; and "Questions for a Sacred Life," 238–42; spiritual, 241–42

death care movement, 241–42

deforestation, 4, 26, 272

Delhi, India, 99
Dellinger, Dave, 193
Deming, Alison, 232
democracy, 166, 167–68, 254
Department of Toxic Substances Control (DTSC), 129–30
Descarte, René, 159
desert, 232, 236
despair, xii, xiv, 214, 233
diaspora, 150, 153
DiCaprio, Leonardo, 199–200
dioxin, 128–29
dissociation, 154
Divine. *See* God; Goddess
Dobzhansky, Theodosius, 145
Dominguez, Joe, 44, 47
Doorway into Light, 242
DTSC. *See* Department of Toxic Substances Control
Duncan, David James, 4
Dur, Gus, 209

E

earth (*see also* nature); anger of, 262–63; and the body, 55, 56, 145, 152, 153, 154; disconnection and dissociation from, 154; and the economic paradigm, 166–67; hiring the class of 2009, 3–4; instructions for living on, 3; as an island, 115, 120; as a living organism, 49, 52; as Mother, 142, 167; movements to save, 4–5, 63–64; new paradigm for living on, 167; our duty to take care of, 142, 167; relationship with, 133, 137, 146–47, 148–49, 150, 152, 170–74, 261; rights of, 167, 168–69; and self-knowledge, 146–47, 148–49; self-regulation of life on, 49–50; stopping the culture that is killing, 132–34; as web of relationships, 171, 172
Earth Day, 84, 168, 227
Earth Democracy, 167–68

EarthFlow Design Works, 61
earthwork, 152, 154–55
eco-anxiety, 58–59, 201, 203
ecological amnesia, 146–47, 149
ecological footprint. *See* carbon footprint
ecology, 193; and the body, 162; as inherently subversive, 250–51
economy: collapse, 27; current paradigm, 166–67; deathly, 132; failed, 5–6; free market, 274; future, 5; and marketplace votes, 77, 78; and nonviolence, 168
ecopsychology, 142
ecosomatics, 151, 152–53, 156
ecotherapy, 201
Eco Village of Ithaca, 269–70
Eden, 210
Edwards, Sarah Anne, 201
ego, 148
Einstein, Albert, 137, 265
11 Hour, The, 200
Emerson, Ralph Waldo, 7
emotional intelligence, 162
End of Nature, The (McKibben), 99
energy conservation. *See* conservation, energy
EnlightenNext Discovery Cycle, 10
environment. *See* earth; nature
Environmental Age, 147
environmental catastrophe, 17 (*see also* global warming); and alienation from the body, 163; causal patterns, 76; and doomerism, 59; how to live in the face of, ix–xv, 18–23, 54–56, 265–66; likened to HIV diagnosis, 53, 54–56; as opportunity, 62, 67; organizations to prevent, 4–5, 63–64, 279 (*see also* activism); and population, 15–17; and predicting outcomes, 79–80, 277, 279–80; questioning thoughts about, 188, 189–90; and a sense of perspective, 55; and a

sense of purpose, 55; as spiritual crisis, 62; stopping the culture that is creating, 63, 64–65, 66, 130–31, 132–34; and the waking-up syndrome, 201–2

environmental movement, 16, 30–31

environmental toxins, 13, 126, 127, 128–29, 226

Essay on the Principle of Population, An (Malthus), 15

Estés, Clarissa Pinkola, 291

Evergreen, 94, 95

evolution, 9, 28, 50, 51, 145, 148

evolutionary psychology, 149

evolutionary spirituality, 10

exercise, 21–22, 163

extinction event. *See* environmental catastrophe

F

Fair Trade movement, 77–78

farmers, 120

farmers' markets, 204, 249

farming, industrial, 66, 166–67

fear: and activism, 69, 227, 229; and the condition of the world, 239; and courage and trust, 80; and decision-making, 58; part of the problem, 62; as a story of a future, 188; and survivalism, 103

Feathers, Susan, 52

Feldenkrais, 164

Feldenkrais, Moshe, 165

Fifth Sacred Thing, The (Starhawk), 230

First Nations, 260, 262

Fisher, Andy, 206

Five Rivers Council, 230

Fog of War, The, 253

food: and carbon footprint, 30, 35, 81; and caring for the earth, 167; and civil disobedience, 66; crisis in participation, 119, 120; ecological and fairness effects of choices about, 79; fundamental for communities, 60; growing (*see* gardens and gardening); industrial control of, 115; local, in eighteenth century, 60; localizing, 59, 66, 119, 248–49; security, 93

Food First: The Institute for Food and Development Policy, 81

food industry, 13, 66, 77, 166–67

football, 161

Ford, 272

ForestEthics, 199

Fort Bragg, CA, 125, 126

For the Common Good (Daly and Cobb), 21

fossil fuel: as adolescent growth hormone, 25, 26; conveniences, 12; diminishing supply of, 25; and food energy, 35; and global-scale alteration, 13, 26, 166; physical labor replaced with, 36; specialization made possible by, 32–33

Frisch, Karl von, 144

"From Mourning into Daybreak" (Simons), 217–24

frugality, 43–44

full catastrophe living, 230

Fuller, Buckminster, 3, 160, 165

future, 8–9, 43–44, 58, 188

Future of Food conference, 115

G

Gadd, Ben, 23

Gaia, 154, 167, 227

Gaia: A New Way of Viewing the Earth (Lovelock), 49

Gailani, Fatima, 209

Gandhi, Mohandas Karamchand, 64, 65–66

Garden Jane, 95

gardens and gardening: benefits of, 35–37, 249; and civil disobedience, 66; community, 35, 104;

gardens and gardening (continued)
learning, 118; most popular pastime
in US, 249; and seed saving, 93–95;
urban, 104–5; and Vicki Robins'
letter from 2030, 42–43
Gates, Bill, 107
Georgia Pacific Corporation, 125–26,
128, 129, 130, 131
Georgia Pacific Corporation mill site,
125, 126–29, 130, 131
Gerry, Father, 257, 258, 259
global warming, 28 (*see also* climate
change; environmental
catastrophe); attitudes toward, 12,
33–34, 50, 129; and blame, 86–87;
and consumer awareness, 272;
fueled by oil, 26; positive
consequences in northern US and
southern Canada of, 247–48; and
social change, 198–99; timing of,
30–31, 33, 213–14 (*see also* tipping
point); and voting, 86
Global Wisdom Council (GWC), 41
God, 4, 8, 28, 210, 243, 244 (*see also*
Goddess); as reality, 188, 190
Godboo, Jerome, 94
Goddess, 24, 244
Goldman, Dana, 75
Gore, Al, xii, 29, 33
Grameen Bank, 107–8
Grandmothers Council. *See*
International Council of Thirteen
Indigenous Grandmothers
gratitude, xiv, 55, 172, 223, 244
Great Barrier Reef, 99
Green, Stephanie (Faith), 78
green architecture movement, 274
Green Belt Movement, 138
green building, 250; carbon-neutral,
275–76; economic benefits of, 272,
274, 276; optimism about, 273; and
Walmart, 271, 272
Green Hunt, 166
Green Party, 21
Greenpeace, 100, 199

green products, 248, 249
Gross National Happiness, 186
Gross National Product, 186
Groupe Danone, 107–8
Guatemala, 86
guilt, 203
GWC. *See* Global Wisdom Council

H

habits, 13–14
hairy rhino, xiv, 183, 184, 187
Haiti, food program for children,
256–58
Hanna, Thomas, 151
Hansen, James, 33
Hartley, Ruskin K., 70
Hartmann, Thom, 46
Havel, Vaclav, 34–35
Hawaii, 116
Hawken, Paul, 7, 63
Hayes, Jane, 95
health and healing, 54, 151, 163, 164,
167 (*see also* spiritual practices)
Hegel, Georg Wilhelm Friedrich, 195
Heinberg, Richard, 206
Heller, Chaia, 196
hiking, 163, 186, 187, 203
Hillman, James, 291
Hilton, Gretel, 116
HIV, 53–54, 56, 76
Hoffmann, Abbie, 15
Hollender, Jeffrey, 110
Holocaust, 18
homesteading, 60, 102–5, 112–13 (*see
also* sustainability)
honeybee, 144
hope, 172, 233, 279–80, 283–85
Horgan, John, xiv, 255
humanity: adolescent identity crisis
of, 24, 25–28, 230; and choice, 8–9,
31, 54, 76, 77, 79, 80, 81, 263; fate
of, 18; as gods, 8–9; relationship

with nature, 51, 55–56, 69–70, 80–81, 137–38, 140–44, 250 (*see also under* earth); responsibility for the future, 9; two things in common for all, 238; uniqueness of, 143
human rights, 4, 166, 169, 254
humility, 79–80
hunger, 4, 76, 80, 167
Hurricane Katrina, 106, 228
hybrid vehicles, 13, 34, 87

I

ice caps, 30, 80, 88, 100, 112, 142
idling vehicles, 82, 83–84
imaginal cells, 265, 266
Imhoff, Marc, 16
Inconvenient Truth, An, xii, xiii, 29
India, 64, 65, 166, 168
indigenous cultures, 153, 156, 173
indigenous mind, 171, 172–74
initiation, 26–27
inquiry, 188–92
inspiration, 4–5, 174
Institute for Circlework, 268
Institute for Social Ecology, 193, 196, 251
integrative medicine. *See* alternative medicine
intelligence, 93, 162, 164
International Association for Ecotherapy, 206
International Council of Thirteen Indigenous Grandmothers, 260, 261, 262, 263–64
Into Stillness (Pallant), 184
intuition, 162
Iroquois, 93
Isho Upanishad, 168
island, 115, 117, 118, 120

J

Jackson, Wes, 117
jaguars, 234, 235

Jamaica, 116–17
Jensen, Derrick, xiv, 134
Judith, Anodea, 28
Juvenal, Decimus Iunius Iuvenalis, 161

K

Kabat-Zinn, Jon, 230–31
Katie, Byron, 85, 190, 192
Kayumari, 261–62, 263, 264
Kenya, 138
Keogh, Martin, 291
khadi, 66
Khan, Hazrat Inayat, 242
Kierkegaard, Søren, 133
Kinesis, 165
kinesthetic intelligence, 162, 164
King, Martin Luther, 63, 64, 65, 156
Kingsolver, Barbara, 14
Knutzen, Erik, xiv, 105
Koch Industries, 129, 130
Kroesen, Kendall, 233, 237
Kunstler, James Howard, 204

L

Lamberton, Jessica, 234, 236
Lamberton, Ken, 237
Lancaster, Rodd, 233, 237
land. *See* earth
landscape. *See* earth
land tenure, 118
Lappé, Anna, 81
Lappé, Frances Moore, xiv, 81
Let's Talk America, 48
Lewis, Samuel, 242
Liberty Media, 106
life: and carbon dioxide, 49–50; conditions conducive to, 5, 6–7; humanity's ignorance of, 51; out of balance, 239; sacredness of, 238–42, 243–44, 261; self-regulating, 49, 50

Life Givers, 261, 263
Linson, Erica, ?
listening, 267
"Localize Enterprise," 60
localizing: and the body, 151; currency, 63; food and everything else, 59–60, 248–49; and khadi, 66
Lopez, Barry, 149, 209, 212
Lord God Bird, 12, 14
Louv, Richard, 206
love: action sustained by, 69; as a force for change, 214; of power versus the power of, 27, 28
Lovelock, James, 49, 52
Lovins, Amory, 160
Luddites, 44

M

Maathai, Wangari, 138, 139
Macy, Joanna, 66, 186, 202, 206
Malaysia, 183, 187
Malthus, Thomas, 15
martial arts, 163
materialism. *See* consumerism
Matrix, The, 201
Mazur, Thaïs, 131
McCallam, Ian, 149
McDonald's, 77
McDonough, William, 271–72, 274
McHugh, Jamie, 56–57
McKibben, Bill, 101, 206
McNamara, Robert, 253
meditation, 163, 186, 187, 203, 204, 266
Mercy Corps, 4
Metta Center for Nonviolence Education, 67
Metzner, Ralph, 206
Michnik, Adam, 34
Millennium Ecosystem Reports, 198
mindfulness, 213–16; and full catastrophe living, 230–31

miracles, 265–67
Mollison, Bill, 113
Moore's Law, 42
Morales Ayma, Juan Evo, 168–69
morels, 11
Mother Earth
Mother Nature. *See* nature
Mother Theresa, 6
mourning, 217–24
Mueller, John, 254
Murdoch, Rupert, 106
Musayev, Lina, 77–78
mycoremediation, 130, 131
Myers, Tom, 165

N

Nagler, Michael, 67
NASA, 16, 33
natural building, 250
natural selection, 8
nature, 19 (*see also* earth); confused relationship with, 142–43; and gratitude, 55; hiring the class of 2009, 7; living in the face of its recession, ix–xv; as medicine, 56, 140–41, 143; relationship to, 51, 55–56, 55–56, 69–70, 80–81, 137–38, 141–42, 250; separation from, 140–41, 142–43; as teacher, 138; viewed in America, 51
Navaho, 22
New Age, 204
New Road Map Foundation, 48
New Zealand, 30, 36
Nobel Peace Prize, 138
Nomad Cohousing, 270–71
nonviolence, 63, 168
North Coast Action, 126–27, 131
Nyland Cohousing community, 270

O

ocelots, 233–34, 235
oil. *See* fossil fuel

"oiligarchy," 26
Okanagans, 152
old growth forests, 197
Oliver, Mary, 5
Oliver Ames High School, 90–92
operation Green Hunt, 166
optimism, 4–5, 160, 253
Orr, David, 206
overpopulation, 15–17

P

Paley, Grace, 193
Pallant, Cheryl, 187
Panther, The (Rilke), 146
paradigm, 166–67
Parham, Opeyemi, 231
past, 43–44
PCBs, 126, 127
peace, 4, 162, 191, 255
"Peace of Wild, Things, The" (Berry), x
"People Power: It's Time to Stop the War Ourselves" (Solnit), 65
permaculture, 111, 151, 204, 205, 233, 270 (*see also* sustainability)
Permaculture Institute (USA), 61, 113
Petrini, Carlo, 248–49
physical education, 160–64
Physicians for Social Responsbility, 227
Pilates, 163
Pilgrim, Agnes Baker, 260, 263
Pinker, Steven, 254
"piti piti n a rive," 256, 258, 259
planets, 49
play, 151, 164
Poland, 34
politics, 122–24, 251 (*see also* activism)
Pollan, Michael, 37, 249
pollution, 26, 76, 272
polychlorinated biphenyls. *See* PCBs

Polynesians, 116, 118
Pomo Indians, 126
population, 16, 21; reduction, 20, 50, 51, 255; regulation of, prior to the human, 49, 50
power, 172–73, 277–79
power shopping, 77
Practice for Living/Living Practice, 174
practices. *See* spiritual practices
Preston, Daniel, 232, 233
Prevatt, Jeneane, xiv
Prevatt, Jyoti Jeneane, 264
psychology, 145, 147, 151, 154, 225

Q

quarencia, la, 149
"Quest for Global Healing," 209–12
"Questions for a Sacred Life" (Bodhi Be), 238–42

R

Rainforest Action Network, 100
rainforests, ix, 80, 142
Rancho El Aribabi, 235–36
Ravindra, Munju, 182
Reagan, Ronald, 34
reality, 188, 190
recycling, 50, 51, 87, 88, 274; at Oliver Ames High School, 90–92
Red Dog, Renee, 232, 233
Rediscovery of North America, The (Lopez), 149
redwood trees, 68–69, 70
re-indigenization, 150, 153–54
relationships, 55–56, 210, 211 (*see also* community)
Reverence for Life, 52
Rich, Adrienne, 4
Rilke, Rainer Maria, 146
Rio Cocóspera, Mexico, 233, 234, 235
ritual, 151, 153–54

rivers, 232, 234, 236
Robin, Vicki, 47–48; letter from 2030, 41–47
Robles, Carlos, 235–36, 237
rock climbing, 163
Rodwin, Scott, 276
Rolf, Ida, 165
Rosencranz, Ann, xiv, 264
Rosenkrantz, Loie, 126, 127
Rubury, Eric, 89
Russia, 277 (*see also* Soviet Union)
Rwandan coffee cooperatives, 77

S

Sabbath, 35
"Sabbaths: VI" (Berry), 283–85
Santa Cruz River, AZ, 232, 234
Santoyo, Larry, 61
Satyagraha, 66
Save the Redwoods League, 69, 70
Schumaker, Peggy, 236
Schweitzer, Albert, 51–52
Search for a Nonviolent Future, The (Nagler), 66
seed saving, 93–95, 112
Seeds of Diversity, 94
self, 146–47, 148–49, 152–53, 162–63, 171
Sense of Wonder, A (Carson), 178
sensori-motor amnesia, 164
Seven Directions Practice, 174
Seventh Generation, 107, 108–9, 110
Sharp, Granville, 5
Shiva, Vandana, 168
SIA. *See* Sky Island Alliance
Silent Spring (Carson), 226
Simon, Vivienne, 216
Simons, Nina, 224
simple living, 248
Simplicity Circle, 203, 204
Simplicity Forum, 48

singing and song, 153–54, 163
Sky Island Alliance (SIA), 234, 235, 236
Slow Food movement, 248–49
Small Planet Institute, 81
Smith, Adam, 32
soccer, 161
social catastrophe, 76, 77, 79–80
social ecology, 194, 250
social revolution, 63, 64–65, 66
solar technology, 35, 189, 273
Solnit, David, 64–65
Solomon, George, 54
soma, 151, 154
Somatic Expression, 56–57
Somatic Movement Therapy, 56
somatics, 153, 160, 162, 163, 164; defined, 151–52; and earthwork, 155; and ecological health, 151
Somé, Malidoma and Subonfu, 291
Soviet Union, 119, 254, 277, 278, 279
spaceship earth, 3, 160
specialization, 32, 33
species, 143–44, 236; marine, 16
Specter, Michael, 30
Spiral Dance, 227
spiration, 174
spiritual awakening, 227, 228; and bodily states, 159; lacking in Americans, 226; and the liberation of human choice from unconsciousness and self concern, 9; and our oil-based adolescence, 26–27; and social revolution, 66
spirituality, earth-based, 227
spiritual practices (*see also* health and healing); contemplation, 27; inquiry, 188–92; listening, 267; meditation, 163, 186, 187, 203, 204, 266; mindfulness, 213–16, 230–31; Tai Chi, 163; walking, 163, 186, 187, 203; witnessing, 186–87; wonder, 177–81; yoga, 155, 163, 178, 186, 196, 203, 204

sports, 161
Starbucks, 78
starfish story, 273
Starhawk, 227, 230
stars, 6, 7
Stonyfield Farm, 107
storytelling, 153–54
Streep, Meryl, 121
stress (see also anxiety); as bodily response, 162; as environmental toxin, 226–27; and intelligence, 93; questioning thoughts that cause, 189; stress taught by, 191
Structural Integration, 164
survivalists, 20, 103
sustainability, 79, 203 (see also homesteading; permaculture); and building (see green building); challenge of, 269–76; and consumer awareness, 272; and corporate consciousness, 107–8; in the desert, 233; and earth rights, 168, 169; good for long-term bottom line, 272; groups, xi, xiii; and homesteading, 60, 102–5, 112–13; moral imperative of, 271–72; moving toward, 204–5; obstacles to, 251; and the tipping point, 276; urban, 60, 105; utilizing the arts to shift toward, 155; in Vermont, 248
Sustainable Seattle, 48
SUV's, 82–83, 84, 86

T

Tai Chi, 163
Takelma Band, Rogue River Indians, 260
Tasmanian Inland Fisheries Department, 111
technology, 153, 166; soft, 153, 156; solar, 35, 189, 273; wind, 13
Terra Foundation, 61
Thirapantu, Chaiwat, 209
Thoreau, Henry David, 177, 273

350.org, 99, 100–101
time, 143
tipping point, xiii, xiv, 65, 80, 213, 276 (see also global warming, timing of)
Titanic, 43
Todd, John, 126
"To Do the Will of God, Come What May" (Walker), 243–44
Tohono O'odham Nation, 232
Tokar, Brian, 251–52
Tolle, Eckhart, 192
topsoil, 26
Toronto Community Garden Network, 95
Transition Whidbey, 48
transportation, 60
trees (see also rainforests); old growth, 197; planted in Kenya, 138–39; redwood, 68–69, 70
trogon, xiv
Trost, Margaret, xiv, 259
trust, 80
Tshering, Lyonpo Ugyen, 209
tsunami, 64, 228
Turning Tide Coalition, 48
Tutu, Desmond, 209
TwoTrees, Kaylynn Sullivan, 174

U

United States, 13, 278
United Students for Fair Trade, 78
Universal Declaration of the Rights of Mother Earth, 168–69
Universe Story, 203
Upanishads, 63
urban homesteading, 60, 102–5

V

Vasudhaiva Kutumbkam, 167
vehicles: hybrid, 13, 34, 87; idling, 82, 83–84

Vermont, 247, 248, 249, 250
Vesaas, Haldis Moren, 139
violence, 76, 168, 191
virtue, 29–30
voluntarism, 63–64
voting, 86, 87–89

W

WAGES. *See* Women's Action to
  Gain Economic Security
Waggoner, David, 42
walk-about, 147
Walker, Alice, xiv, 244
walking, 163, 186, 187, 203
Walla, Nala, 157
Wall Street, 166
Walmart, 18, 106; and green building,
  271, 272
warfare, 26, 80, 191; decline of, 254;
  fatalism about, 253–55; protesting,
  63, 93; solving, 253, 254, 255
water, 4, 60, 167, 272; and
  community, 236; as holy, 236–37
wealth, 166, 167, 186
Wedgwood, Josiah, 5
What If? Foundation, 258, 259
Whitman, Walt, 56
wilderness, 21–22, 140
Wilderness Foundation Africa, 149
wildness, 140, 177
Williams, Betty, 209

Williams, Jody, 209
Wilson, Edward O., 255
wisdom traditions, 203
witnessing, 186–87
Wolf, Cynthia, 235
women, 76, 222, 255
Women's Action to Gain Economic
  Security (WAGES), 109
wonder, 141, 177–81
woodpeckers: ivory-billed, 12, 14;
  pileated, 11–12
Work, The, 192
workaholism, 226–27
World Future Council, 169
World Trade Center, 19, 119, 266
World War II, 18, 36, 277, 278
Worm, Boris, 16
writing, 184, 186, 187

Y

Yeats, William Butler, 147
YES! Magazine, 63, 65
yoga, 163, 186, 196, 203, 204; and
  earthwork, 155; and wonder, 178
Your Money or Your Life (Robin and
  Dominguez), 46

Z

Zinn, Howard, 280
zone zero, 151

About the Editor

Martin Keogh is the founder of The Dancing Ground, an organization that produces conferences and symposia on gender, race, and mythology. He has produced and taught with such bestselling authors as Joseph Campbell, James Hillman, Clarissa Pinkola Estés, Robert Bly, Coleman Barks, Malidoma and Subonfu Somé, and many others. One emphasis of Dancing Ground events is the need for actively engaging in the world in order to fully enter life.

A Fulbright Senior Specialist, Keogh is also the author of *As Much Time as It Takes: A Guide for the Bereaved, Their Family and Friends.* For his three-decade contribution to the development of the dance form Contact Improvisation, he has been listed in *Who's Who in the World.* He has facilitated groups and teacher's conferences in twenty-seven countries on five continents. His approach to writing and teaching are the same—allowing the body to provide the imagery that viscerally communicates what the body knows. His essays have been translated into eight languages.

For more information, visit his Web site at www.martinkeogh.com.

Permissions and Copyrights